To LEO
CONTESE

AVOID OPINI[ON]
STICK TO THE FACTS
TRIM DOWN TO THE
SCIENCE

DR KARL'S
SHORT BACK & SCIENCE

14/10/16

ALSO BY

DR KARL KRUSZELNICKI

Curious & Curiouser

Brain Food

50 Shades of Grey Matter

Game of Knowns

House of Karls

Dinosaurs Aren't Dead

Dr Karl's Big Book of Science Stuff (and Nonsense)

Dr Karl's Even Bigger Book of Science Stuff (and Nonsense)

Dr Karl's Biggest Book of Science Stuff (and Nonsense)

DR KARL'S
BOOKS SINCE 1985
SHORT BACK & SCIENCE

DR KARL KRUSZELNICKI

Pan Macmillan Australia

First published 2015 in Macmillan by Pan Macmillan Australia Pty Ltd
This Macmillan edition published in 2016 by Pan Macmillan Australia Pty Ltd
1 Market Street, Sydney, New South Wales, Australia, 2000

Copyright © Karl S. Kruszelnicki Pty Ltd 2015

The moral right of the author has been asserted.

All rights reserved. No part of this book may be reproduced or transmitted by any person or entity (including Google, Amazon or similar organisations), in any form or by any means, electronic or mechanical, including photocopying, recording, scanning or by any information storage and retrieval system, without prior permission in writing from the publisher.

Cataloguing-in-Publication entry is available
from the National Library of Australia
http://catalogue.nla.gov.au

Cover, internal design and typeset by Xou Creative, www.xou.com.au

Cover photograph of Dr Karl by Mel Koutchavlis
Other cover images: iStock

Internal illustrations by Roy Chen, Xou Creative

Extracts from Letters From Mesopotamia: Official, Business, and Private Letters on Clay Tablets from Two Millennia by A. L. Oppenheim reprinted courtesy of the Oriental Institute of the University of Chicago.

Printed by McPherson's Printing Group

The author and the publisher have made every effort to contact copyright holders for material used in this book. Any person or organisation that may have been overlooked should contact the publisher.

The publishers and their respective employees or agents will not accept responsibility for injuries or damage occasioned to any person as a result of participation in the activities described in this book. It is recommended that individually tailored advice is sought from your healthcare professional.

Papers used by Pan Macmillan Australia Pty Ltd are natural, recyclable products made from wood grown in sustainable forests. The manufacturing processes conform to the environmental regulations of the country of origin.

I dedicate this book to the Next Generation.

As described by the Flynn Effect, they are nine IQ points smarter than their parents. They are well equipped to deal with what future decades will bring. Personally, I'm an optimist, and remain ever hopeful that things keep getting better – or that we all genetically engineer ourselves into clouds of iron vapour, with a mass of 50 kilograms, intelligent, the size of a planet, and floating through space (thanks, Freeman Dyson).

TABLE OF CONTENTS

01. RADIOACTIVE YOU 1

02. SLIPPERY BANANA PEELS 10

03. BREATHE IN SPACE 18

04. A HOLE THROUGH THE EARTH 24

05. QUANTUM LIFE 32

06. NAKED MOLE RATS DON'T GET CANCER? 46

07. SYNTHETIC SHIRTS STINK 54

08. BLUE DRESS ILLUSION 56

09. BRONTOSAURUS IS BACK 64

10. STARS ARE POINTY 74

11. CORN PASSES STRAIGHT THROUGH 82

12. NAKED IN THE ANTARCTIC 84

13. THE KILLER DRILLER 92

14. SURPRISING SUN 106

15. SATURATED FATS 114

16. COCONUT OIL AND WATER 116

17. OLDEST KNOWN COMPLAINT LETTER 124

18. BREAKING THE SEAL 132

19. WINE GLASS SHAPE AFFECTS FLAVOUR 138

20. MICROMORT 142

21. GREAT BARRIER REEF 146

22. ANTIOXIDANTS & SNAKE OIL 160

23. (NOT) NEUTRAL NIGHTLY NEWS! 170

24. HOLE HEARTED 172

25. COWS MAKE FLAVOURED MILK 178

26. COCONUT WATER VS BLOOD 184

27. WEIGHT OF A CLOUD 192

28. VEINS ARE NOT BLUE 194

29. BACTERIOPHAGE & L'ORÉAL 202

30. EARWAX & ARMPIT SWEAT 210

31. MICE ON WHEELS 216

32. TOILET SEAT GAVE SOUND RANGING 222

33. ANAESTHETICS DOSE 234

34. EXPENSIVE TEACUP 240

35. FAT: WHERE DOES IT GO? 244

36. EARTH'S SPIN AXIS IS SHIFTING 252

37. BOUNCING BATTERIES 258

REFERENCES 269

ACKNOWLEDGMENTS 287

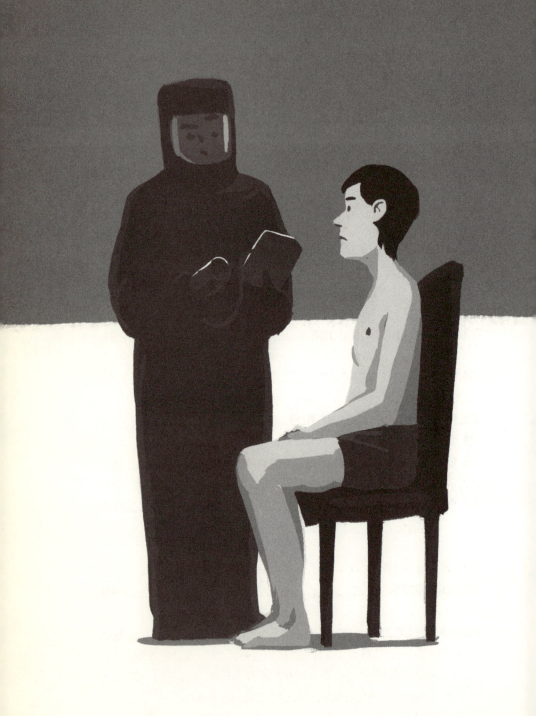

01

RADIO-ACTIVE YOU

EACH OF US IS BATHED IN AN INVISIBLE OCEAN OF RADIOACTIVE PARTICLES AND ENERGIES – ALPHA AND BETA PARTICLES, GAMMA RAYS, X-RAYS, NEUTRINOS, MUONS, ETC. IN THE VAST MAJORITY OF CASES, THIS RADIATION DOES US NO HARM. THAT'S GOOD, BECAUSE YOU CAN NEVER GET AWAY FROM IT. EVERY ONE OF US IS EACH PERSONALLY EMITTING RADIOACTIVITY – 24 HOURS A DAY.

WITHOUT MEASUREMENT THERE IS NO KNOWLEDGE

The mathematical physicist Lord Kelvin once declared, "I often say that when you can measure what you are speaking about, and express it in numbers, you know something about it; but when you cannot express it in numbers, your knowledge is of a meagre and unsatisfactory kind; it may be the

RADIOACTIVITY UNITS

First, let's start with how we measure radioactivity. By way of comparison, we measure human weight in straightforward units such as kilograms (or pounds), and human height in metres (or feet and inches). But measuring "radioactivity" is a little confusing, because radioactivity is complex.

So let me show you how complex it is by giving an example from martial arts. Consider a Mixed Martial Arts athlete throwing punches and kicks in all directions. You can count how many blows she throws, the type of blows, how many of them connect, and finally, what kind of damage they cause. In the Land of Radioactivity, there is a different unit for each of these measurements (sorry).

Let's assume that most of the blows miss you. Only some of them land on you. Some might be hard and land like a sledgehammer, while others might be as soft as a feather.

Moving on, think about where the blows land.

If you suffer 50 little punches on your arms and legs, you wouldn't suffer any permanent damage. But if those 50 little punches were all to one delicate eyeball, you might well end up permanently blind. When we talk about Radiation Landing On Your Body, different parts of the

beginning of knowledge, but you have scarcely, in your thoughts, advanced to the stage of *science*, whatever the matter may be."

In other words, in the Field of Science, if you can't measure something, your understanding is inadequate.

body are affected differently. Your bone marrow is much more sensitive to radiation than your big toe.

Finally, let's consider the type of radiation. Matching energy for energy, alpha particles do 10 to 20 times as much damage as gamma rays.

So let's go straight to a unit that incorporates all the factors I just mentioned – how much lands on you, where on your body it lands, and the type of radiation.

This unit of Radioactivity is called the Sievert (Sv). Sieverts are a rather large unit, so we use milliSieverts (mSv). The Americans have remained steadfastly non-metric and use the Roentgen Equivalent in Man (REM). In general, your lifetime risk of developing a fatal cancer is about one in four. Getting exposed to 10 mSv means you now have (on average) a roughly one in 2000 chance of eventually growing an extra cancer. (And it also increases your lifetime risk to about 1 in 3.995.)

TYPES OF RADIATION, PART 1: IONISING

The most common types of harmful radiation are alpha, beta and gamma. When they "smash" into an atom in your body (or interact with it via electromagnetic effects), they rip away one or more of that atom's electrons. The "atom" has been turned into an "ion" – so these types of radiation are called "ionising radiation". These ions then cause damage at a cellular level.

Alpha and beta radiation are particles. They don't penetrate the human body very well. Alpha particles can be stopped by a tissue paper, or the outer layers of your skin (which is about 40 microns, or a few cells, thick).

Alpha particles are basically the core (or nucleus) of a helium atom without the surrounding two electrons. An alpha particle is just a tiny "heavy" ball of two protons and two neutrons.

A beta particle is a high-speed electron (or a positron, an anti-matter particle, which is an electron with a positive charge). Beta particles penetrate the body better than alpha particles, but not as well as gamma radiation. They can be stopped by a few millimetres of aluminium.

Gamma radiation is not a particle. It's high-energy electromagnetic radiation. Gamma radiation is less ionising than alpha or beta particles, but penetrates more deeply into the body. You need serious shielding, such as lead, to stop gamma radiation.

BREAKDOWN OF RADIATION DOSE – BIG DUDES

In the USA (for which we have good figures), the average annual dose of radioactivity is about 3.6 mSv. That's equivalent to about 180 chest X-rays, or 18 return flights to London.

In the USA, some 2 mSv of that dose comes from radioactive radon gas. (For Australia, which has different geology, the figure is about 0.2 mSv). This radioactive radon usually comes from granite. Granite naturally carries tiny amounts of uranium – which decays into radon. When radon decays, it emits alpha and beta particles, and gamma rays. There are many locations in the USA where granite is near the surface. Furthermore, if you have granite kitchen benches, you pick up a little extra radon in your kitchen.

Next are medical X-rays, giving the average American about 0.53 mSv per year. (The latest Australian figures are about 1.7 mSv per year.)

The next major background Radiation Contributor is the "world" inside our skin. Atoms of radioactive potassium-40, carbon-14 and the like get breathed in, eaten or drunk, and then get incorporated into our bodies. Their radioactive decay gives us 0.4 mSv per year.

RANKING RADIATION RISK

SOURCE	DOSE (MSV)	NO. OF CHEST X-RAYS
Head MRI scan	0	0
Live next to nuclear reactor, 1 year	0.0005	0.025
Chest X-ray	0.02	1
Sydney–London return flight	0.2	10
Live in Sydney, 1 year	2.5	125
Live in Tibet, 1 year	3.2	160
Bone scan (Technitium-99m)	4	200
Abdomen CT scan	5–10	250–500
Cardiac catheterisation	>10	>500

BREAKDOWN OF RADIATION DOSE – LITTLE DUDES

Furthermore, the natural world around us sprays us with 0.28 mSv per year. Clothes, paper, concrete and yes, food such as bananas and brazil nuts, are all slightly radioactive.

Those mysterious invaders from outer space, cosmic rays, give us another 0.27 mSv per year. They smash into atoms of nitrogen and oxygen in our upper atmosphere, which then emit muons. (You can think of muons as very heavy electrons – about 200 times heavier.)

But as muons head down towards the ground, they get re-absorbed by the atmosphere (and don't give off any more harmful radiation). So this particular type of radiation decreases as you get closer to sea level. Therefore, it increases with altitude – an extra 0.01 mSv per year for each 30 metres. You receive your lowest possible cosmic radiation dose at sea level, but you get significantly more in the mountains – an extra 1 mSv per year if you live at 3000 metres. You pick up radiation very quickly – about 0.01 mSv per two hours – with a jet flight, because you are so much higher, up around 10,000 metres. On the other hand, most of us don't fly thousands of hours each year.

Consumer products, such as battery-powered smoke detectors, come next at about 0.1 mSv per year. This is because of their internal radioactive americium-241, which emits alpha particles. (By the way, most mains-powered smoke detectors are not radioactive. They are usually photo-electric in nature. Not only are they more effective at registering smouldering fires, they also generate fewer false alarms.)

Also giving the average American another 0.01 mSv each year is atmospheric radioactivity from past nuclear bomb blasts. This figure is a little controversial. Nuclear bombs were exploded in the atmosphere from 1945 until 1980. Underground nuclear tests, which normally don't dump radioactivity into the atmosphere, finished with India and Pakistan in 1998. However, North Korea announced it had tested a "nuke" in 2013.

Another 0.0005 to 0.01 mSv per year comes from US nuclear power plants. While the USA has about 100 commercial nuclear power plants, Australia has only one nuclear reactor – but it's not a huge power reactor. Instead, it's a tiny reactor that is used for research. This reactor also supplies us with essential medical radioactive isotopes, such as technitium-99m. (On average, during their entire life, each Australian will have one diagnostic or therapeutic medical procedure involving a radioisotope.)

RISK OF EVERYDAY ACTIVITIES

Various activities have different risks, which change depending on the location and the time. For example, growing crops is very risky if the countryside has many unexploded mines scattered around.

So in that spirit, here are the risks for the specific location of the United Kingdom, back in 1987.

ACTIVITY	RISK OF PERSON DYING IN A CALENDAR YEAR
Smoke 10 cigarettes per day	1 in 200
Influenza	1 in 500
Road accident	1 in 8000
Radiation, 1 mSv	1 in 20,000
Accident at home	1 in 26,000
Accident at work	1 in 43,000
Hit by lightning	1 in 10,000,000

RADIOACTIVE CIGARETTES

Don't forget radioactive ciggies (which most of us don't smoke). In my 32nd book, *50 Shades of Grey Matter*, I wrote how two packets of cigarettes give you as much radiation as one chest X-ray.

(The radiation comes from polonium-210.) If you smoke two packs of cigarettes per day, that's equivalent to about one chest X-ray, or 0.02 mSv per day. Over a year, that works out to about 730 mSv.

That is just one of the many, many reasons why smoking is so bad for you.

BREAKDOWN OF RADIATION DOSE – BELOVED DUDES

Finally, there's your bestie and beloved, sleeping peacefully beside you for one third of each day. Human bodies are mini powerhouses of radioactivity, emitting some 0.4 mSv each year. We are all aglow, both with love and radioactivity.

Most of your partner's radiation is trapped by their flesh.

But a tiny amount escapes to land on you. How much? Just one tiny 0.01 mSv each year.

That particular risk is one that I'm very happy to wear, and share . . .

TYPES OF RADIATION, PART 2: NON-IONISING

There are many types of electromagnetic radiation that do not create ions when they collide with your body. So they are called Non-Ionising Radiation.

They include the electromagnetic radiation from visible light and infrared, radio waves, TV waves, microwaves and mobile phones.

They have never been proven to cause cancers.

02

SLIPPERY BANANA PEELS

THEY'RE YELLOW, RADIOACTIVE, HIGH IN POTASSIUM AND VERY SLIPPERY. (OR ARE THEY? SOME OF WHAT YOU JUST READ IS WRONG. SO READ ON.)

Yes, they're bananas and we have just found out why they are so slippery. It's because of carbs and proteins that are trapped inside tiny bumps on the inner layer of the banana skin.

POTASSIUM IS ESSENTIAL FOR ALL LIVING CREATURES

In fact, the name potassium comes from "potash" – the ashes of plants. So growing lots of plants will deplete the soil of potassium. It turns out that 95 per cent of the world's chemical production of potassium is used in agricultural fertilisers.

Cells accumulate potassium.

All animals cells have the famous

YELLOW?

OK, most of the bananas we see are indeed yellow.

But there are two caveats.

First, if you go looking around the world, you'll find that bananas usually come in other colours. When ripe, a banana can be green, red, brown or even purple. But yes, the bananas which we Westerners mostly see are yellow – mainly the Cavendish and Gros Michel varieties.

Second, the yellow colour we see is actually more intense than "natural", thanks to the artificial ripening process. Many bananas are picked while immature, and then stored and transported at around 13–15°C. Once they arrive close to their final destination, they are held at 17°C and exposed to the ripening gas ethylene.

So, bananas are yellow? Yes, quite often in wealthy Western countries.

RADIOACTIVE?

Yes, bananas are slightly radioactive. In fact, they are even used as the basis for an informal Unit of Radioactivity – the Banana Equivalent Dose (or BED). It was invented to give the general public a handle on radioactivity.

Bananas contain potassium, the second least dense metal (after lithium). The "average" 70 kilogram human carries about 150 grams of potassium. It occurs naturally as three isotopes. Potassium-39 (93.26 per cent)

(to medical students, at least) Sodium-Potassium Pump. For every two potassium ions it pumps into the cells, it also pumps three sodium ions out of the cell. This maintains the electrochemical gradient across our cell membranes – where the outside is negative by 70–80 milliVolts, relative to the inside of the cell. This Sodium-Potassium Pump uses about 20 per cent of the energy your cells need.

and potassium-41 (6.73 per cent) are stable.

But potassium-40 (0.0117 per cent, or roughly just one out of every 8550 potassium atoms) is unstable – it's radioactive. It has a half-life of 1.25 billion years (about nine per cent of the age of the Universe). It decays to either stable calcium-40 or stable argon-40. (This nuclear decay reaction allows us to "date" the age of rocks.)

In your body, about 4400 atoms of radioactive potassium-40 decay each second. It's our major source of natural human radioactivity, more than we receive from the decay of carbon-14. Mind you, the potassium-40 atoms don't specifically accumulate in our bodies in preference to "regular" potassium atoms – all potassium atoms get treated the same way. Potassium is very dynamic in the human metabolism, continually entering and leaving our bodies.

The actual radiation dose from a single banana (the BED) is about 0.1 microSievert (µSv). This is about one per cent of your average daily exposure to radiation (see "Radioactive You", page 1). A CT scan of the chest delivers seven milliSievert (7000 µSv, or 70,000 BED). The lethal dose of radiation is about 35 million BED (or 3500 mSv).

The USA has set up a network of extremely sensitive instruments at Radiation Portal Monitors to detect smuggling of nuclear materials. The radioactivity from a truck fully loaded with bananas is great enough to activate these Monitors.

But should you be worried about eating bananas? No, their radiation dose is so low that it's not harmful.

HIGH IN POTASSIUM?

Bananas do in fact contain potassium – about 358 milligrams per 100 grams of banana. But they are not especially rich in potassium.

The United States Department of Agriculture (USDA) publishes a National Nutrient Database for Standard Reference. Out of some 8019 food items that were rated for their potassium content, bananas came in around 1600th place.

Right at the top was Cream of Tartar, with 16,500 milligrams per 100 grams. But that's not a "food item" that you can eat like an apple (120 milligrams per 100 grams) or a potato (650 milligrams per 100 grams).

Pumpkin seeds (919 milligrams per 100 grams), yam (495), parsley (554), dried apricots (443), almonds (733), bamboo shoots (533), avocados (507), soybeans (515), wheatbran (1182), kidney beans (419), lima beans (508), sunflower seeds (850), cashew nuts (660), brazil nuts (659), and Spanish peanuts (744) are all higher in potassium than bananas. (I am waiting for potatoes to be spruiked as the next "Superfood".)

On the other hand, bananas are easy to peel and eat.

SLIPPERY?

Let's look at the slipperiness of a banana peel.

Unless you work at a big fruit market you probably haven't seen a real person slip on a banana peel – let alone a bride! But one fruitologist told me he had seen a forklift truck spinning its wheels on banana peels, totally unable to get any traction.

Of course, in cartoon-land, banana skins are very slippery indeed. Bugs Bunny, for example, tosses the skin of a peeled banana onto the floor, and then Elmer Fudd steps on it, windmills his legs and falls.

The wascally wabbit wins once again.

What's odd is that we all "know" that the banana peel is slippery, even though most of us have never seen it in action.

Back in 2012, some "tribologists" in Japan – Kiyoshi Mabuchi and colleagues at Kitasato University in Minato – decided to measure just how slippery a banana peel was.

TRIBOLOGY

Tribology is the study of surfaces that move relative to each other. Tribology looks at "friction", as well as "lubrication" and "wear". The word comes from the Greek root "trib", meaning "I rub".

The field of tribology really began in the mid-1960s. It was started off by the physicist David Tabor, and the lubrication expert Peter Jost. But humans have been using the principles of tribology for much longer than that. Nearly 5000 years ago, the ancient Egyptians used jugs of lubricant to move giant stones to build the pyramids. In the late 1400s, Leonardo da Vinci devised the first Laws of Friction.

Today, friction costs developed countries one to two per cent of their GDP. Future areas of tribology include nanotribology (dealing with very small particles), biotribology and green tribology.

Friction is very complicated and depends on many factors. One is the speed at which the two surfaces are moving, relative to each other. For example, static friction is quite different from dynamic, or moving, friction. You might have noticed that it's hard to get something moving, but once it is moving, you need much less effort to keep it going. Other factors include the materials that the surfaces are made from, the properties of any lubricants, and so on.

One easy way to measure friction is the so-called "Coefficient of Friction" or "Frictional Coefficient".

Suppose you have a block of leather, weighing 1000 grams, resting on a wooden surface. To move it, you have to push it with an equivalent weight of 350 grams. The Frictional Coefficient of leather-on-wood is

BANANA REPUBLIC

The term "Banana Republic" originally referred to countries in Central America (including Costa Rica, Honduras and Panama) that were systematically manipulated for decades by multinational food companies such as the United Fruit Company of Boston. Today, it refers to any nation that is manipulated by corporations.

These powerful companies were "vertically integrated". They totally controlled all aspects of banana production – growing, processing, shipping and marketing. The monopolistic companies used their massive economic and political

350 divided by 1000, which works out to 0.35. For rubber sliding on concrete, it's about 1.02. The kings of low friction are ice-on-ice and metal-on-ice (around 0.02) and the articular cartilage in your joints (0.01).

BANANA TRIBOLOGY

The Japanese tribologists measured the frictional coefficient of banana peel on a linoleum surface as 0.07. This was amazingly low – better than metal-on-metal lubricated by oil.

Under the microscope, the researchers found that there were little tiny bumps, or follicles, on the inner banana skin – the surface that touches the delicious fruit. When your foot puts some pressure on these follicles, the follicles release polysaccharides (which are carbohydrates) and protein. These then unite (or combine) to form a homogeneous gel. Thanks to the bumpy nature of the skin, this homogeneous gel gets trapped between the banana and the floor, providing a wonderfully low friction state.

Of course, being thorough tribologists, they measured the frictional coefficient for a few different fruits. Apple peel was not so slippery at 0.1, while tangerine and citron (which is an Asian fruit similar to a lemon) each came in around 0.2.

For their groundbreaking research, the Japanese tribologists won the power to create what the economists call an "enclave economy". An enclave economy is separate from the rest of the country. It is self-sufficient, exports practically all that it produces – and pays virtually no tax to the host country.

The country suffers, while the fruit company gets very wealthy indeed.

light-hearted IgNobel Prize in Physics in 2014. (I won the 2002 IgNobel Prize in Interdisciplinary Science, for my "groundbreaking" research into Belly Button Fluff, and why it is almost always blue. See my 20th book, *Q&A with Dr K*.)

And their Ethics Approval came with the proviso that no cartoon wabbits were harmed or injured during the course of their study – but they were unable to say the same for Elmer Fudds...

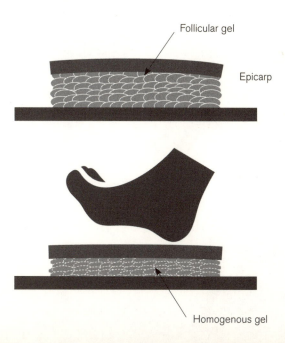

03

BREATHE IN SPACE

ACCORDING TO THE ADVERTISING FOR THE 1979 MOVIE *ALIEN*, "IN SPACE NO ONE CAN HEAR YOU SCREAM." THIS MAKES SENSE, BECAUSE THERE'S NO AIR IN SPACE.

But inside the International Space Station (ISS), the American astronauts and Russian cosmonauts breathe air almost identical to the stuff down here on planet Earth. Where do they get their oxygen?

It turns out they get it by "splitting" H_2O with electricity. It's tricky – or as physicists like to say, "non-trivial".

INTERNATIONAL SPACE STATION

Since 2 November 2000, the ISS has been continually inhabited by humans from some 15 countries.

The ISS is huge – easily visible with the naked eye from Earth. It weighs about 450 tonnes. It's about 108 metres wide (about two Olympic swimming pools), about 73 metres long and about 20 metres high. It orbits the Earth roughly every 93 minutes, travelling at around 27,600 kilometres per hour. It's surprisingly close to us in its Low Earth Orbit – only some 330 to 435 kilometres above the ground. While my home solar cells generate only 4.5 kW, the ISS's massive solar cells generate around 131 kW – but only for slightly more than half of each orbit, while the ISS is in sunlight. When it's in the Earth's shadow (some 35 minutes of each orbit), electricity comes from its rechargeable nickel-hydrogen batteries.

It's this electricity that splits water into hydrogen and oxygen:

$$2\,H_2O \rightarrow 2\,H_2 + O_2$$

There are two Modules that do this. The Russian one is called "Elektron" while the American one is called the "Oxygen Generating System". While the oxygen has always been recycled into the atmosphere of the ISS, the hydrogen used to be dumped overboard. However, it is now combined with carbon dioxide (which is breathed out by the inhabitants and then captured by CO_2 scrubbers) to make water and methane.

This happens in a closed unit:

$$4\,H_2 + CO_2 \rightarrow 2\,H_2O + CH_4$$

The water is then recycled into hydrogen and oxygen.

Quite separately, human "stinks", such as methane from farts, and ammonia from sweat, are removed from the air by activated charcoal filters.

Overall, electrolysis of water provides about 5.4 kilograms of oxygen per day for four people. However, it can be ramped up to nine kilograms when there are six inhabitants.

Of course, there are backup oxygen systems. First, there are Solid Fuel Oxygen Generator canisters, similar to those on commercial jets. Each canister of lithium perchlorate will provide enough oxygen for one person for one day. In addition, there are also some emergency cylinders of bottled compressed oxygen.

It's amazing how complicated it is to copy what plants do for us for free.

SUFFOCATE IN CARBON DIOXIDE

When they go to sleep, the inhabitants of the ISS have to make sure that they are in a well-ventilated area. If there is not good airflow, the carbon dioxide they breathe out will accumulate around their heads.

On Earth, the expired air from our lungs is usually warmer than the ambient air. When you combine this difference in temperature with Archimedes' Principle, the warm air rises and the cooler air falls.

But this rise-or-fall does not happen in the microgravity of space. Before the spacegoers discovered this, they would sometimes wake up gasping for oxygen.

Better to wake up gasping than to not wake up at all.

URINE TO OXYGEN?

The water recovery systems on the ISS catch water from showers, sinks, and water vapour in the air, and feed it into the oxygen generators.

Urine is also processed.

Originally, the systems were supposed to be able to recover 85 per cent of the water in the collected urine. But the microgravity conditions of the ISS (See "Space, the Hostile Frontier" in my 36th book, *House of Karls*)
loss of bone density in the spacegoers, which leads to very high calcium levels in their urine. This reduced the efficiency of the water recovery systems to just 70 per cent.

04

A HOLE THROUGH THE EARTH

SUPPOSE YOU DRILLED A HOLE RIGHT THROUGH THE CENTRE OF THE EARTH AND OUT TO THE OTHER SIDE – AND THEN YOU JUMPED INTO THAT HOLE. WHAT WOULD HAPPEN?

The answer to this beautiful question used to be that you would pop out on the other side of the Earth 42 minutes and 11 seconds later. But a bit of fact-finding and fine-tuning now tells us that it would be closer to 38 minutes and 11 seconds. (Yup, 38 minutes to zip right through the middle of the Earth and out to the other side.)

A bunch of such tunnels drilled straight from one city to another would make a great energy-free transport system.

HISTORY

One of the earliest mentions of the concept of falling straight through the Earth appeared in the French magazine *La Nature* in 1883 with the title "De Paris á Rio de Janeiro en 42'11". It then reappeared in an 1898 engineering textbook, and by the 1930s had become slightly more widely known as a junior-year engineering and physics problem. In 1966, physicist Paul Cooper took a fresh and deeper look at the concept.

FOUR ASSUMPTIONS

Let's start with a few assumptions.

First, let's assume you could actually build this tunnel. You would need both the drilling machines, and the walls of the tunnel, to be made of some incredibly exotic material such as the fictional "Unobtainium". As you drill down towards the centre of the Earth, you would first pass through up to 100 kilometres of solid crust, followed by a few thousand kilometres of molten rock (the earth's "mantle"). At 2900 kilometres down you would plunge into some 2200 kilometres of liquid iron (the "outer core"), and then run smack into about 1300 kilometres of solid iron (the "inner core"). At the very centre, some 6378 kilometres down, the temperature would be hotter than the surface of the sun. Only Unobtanium could withstand this truly hostile environment.

Second, let us make believe there is no air in the tunnel. Obviously,

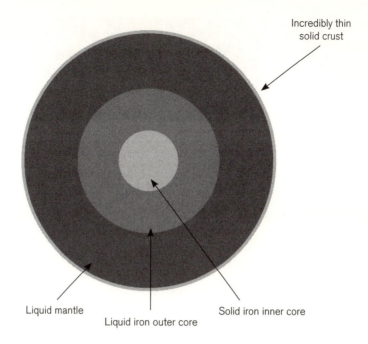

if air was present, the wind resistance would slow you down. But more importantly, once you've fallen several hundred kilometres down the vertical tunnel, the weight of the air above would generate enormous pressure. How much pressure? So much that the air a few hundred kilometres down would have turned from a gas to a liquid and finally a solid (depending on the temperature). You couldn't fall through that. To solve this, let's have a vacuum.

Third, let's blissfully ignore the fact that the Earth is rotating. On the surface of the Earth at the equator, you are moving from West to East at about 1700 kilometres per hour. That's your "sideways velocity", but you don't notice it because the ground is moving at the same speed. However, this sideways velocity at the very centre of the Earth is zero. So when you reached the centre of the Earth, you would be grinding against the side of the tunnel with that same sideways velocity of about 1700 kilometres per hour. So let's ignore rotation.

Fourth, let's assume that the Earth has a constant and average density of about 5.514 tonnes per cubic metre.

WHAT ABOUT OTHER PLANETS?

For Mars (the red planet, next one out from the Sun past us), the one-way transit time through the planet would be 49 minutes. I guess this is because of its lower gravity.

Our Moon? 53 minutes' one-way transit. But it would be easier to make the tunnels there – no air and no molten core.

LET'S JUMP

So now (in our imaginary "perfect" world) that we have our airless, frictionless tunnel, let's jump in.

In the first half of our trip, as we head for the core, we swap gravitational potential energy for velocity. The entire mass of the Earth is under us, pulling on us with its gravity. By the time we reach the dead centre of the Earth, we are moving at about 28,000 kilometres per hour.

If we could magically come to an instantaneous dead halt, we would just float, free of gravity. The gravitational pull of the Earth all around us would be effectively zero. Directly ahead of us would be the mass of half the Earth – but that would be exactly cancelled out by the mass of the other half of the Earth that was behind us.

But once we're reached the core, we have tremendous velocity, and so we keep moving.

In the second half of our journey, we swap velocity for gravitational potential energy. Indeed, as we journey upwards and move further from the core, there is an ever-increasing amount of our planet behind us. So its ever-increasing gravity slows us down. But the amount of energy we picked up on the way down is exactly what we need to propel us all the way upward against the pull of gravity.

After a total time of 42 minutes and 11 seconds we appear on the other side of the Earth. If we grab onto the side of the tunnel, we'll be able to

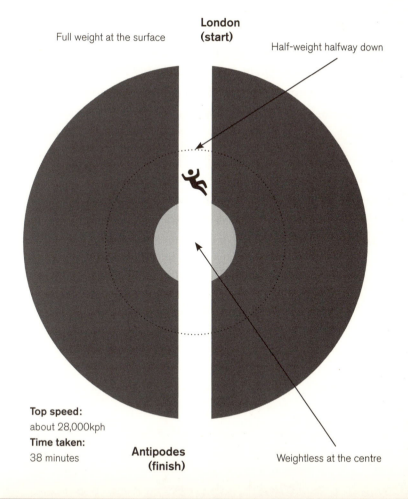

Full weight at the surface

London (start)

Half-weight halfway down

Weightless at the centre

Antipodes (finish)

Top speed: about 28,000kph
Time taken: 38 minutes

A Hole Through The Earth < 29

haul ourselves out. But if we don't, we'll return to our starting point another 42 minutes and 11 seconds later. Like a pendulum, we would oscillate back and forth indefinitely, for ever and ever and ever . . . (The Physicists call this Simple Harmonic Motion.)

ENERGY-FREE TRAVEL

Back in 1966, Paul Cooper wondered what would happen if you drilled straight tunnels between various locations, e.g., Sydney and London, Sydney and Melbourne, Sydney and the Taj Mahal, etc.

He showed that the time for a trip in a straight tunnel, from anywhere on the surface of the Earth to anywhere else, was also 42 minutes.

He wrote, "By criss-crossing our planet with frictionless subterranean passages, man could achieve rapid intercontinental travel comparable in transit time with that of a space vehicle, but with no expenditure of energy . . ." An extra advantage of this transportation system, over and above needing no energy to run it, was the lack of timetables. Cooper proposed, ". . . the world's cities are linked . . . the departure time is universally on the hour and the arrival time 42 minutes later." Think how much you would save by not having timetables!

Cooper proposed that we could now look downward for our future transportation. Ignore space and the sky – ride the Deep Underground instead.

FINE-TUNING TIME

In 2015, Alexander Klotz, a student from McGill University in Canada, took another look at this problem.

He quickly realised that the density of the Earth is not constant. It's less at the surface, and greatest at the core. He used the Preliminary Reference Earth Model – a geological model of our planet that is based on more recent seismic data than Paul Cooper had available. The density of our planet is about one tonne per cubic metre at the surface, but rockets to 13 tonnes per cubic metre at the core. Furthermore, about halfway down to the core, there's a rather dramatic jump in density near where the ball of liquid iron begins.

So when Mr Klotz carried out the calculations with the new data, it turned out that the time to fall through a hole in the Earth is about 38 minutes and 11 seconds – four minutes quicker.

Mr Klotz also worked out that different straight-line trips would take different times. This is due to the varying densities of the Earth along the way. But the time difference isn't great – 42 minutes for short trips, and up to 38 minutes for long trips. This is totally counter-intuitive – long time for short trips, short time for long trips. So you could still do without timetables (the mathematics is a little complicated).

Mr Klotz begins his third paragraph with the words, "it is unlikely that such a tunnel will be excavated in the near future". Sure, the tunnel through the Earth is not shovel-ready, but *near* future" – does he know something the rest of us don't?

Only the passage of time will tell . . .

05 the Wave f()

QUANTUM LIFE

THE WORD "QUANTUM" IS BANDIED AROUND AN AWFUL LOT – ESPECIALLY WHEN IT COMES TO THE HARD SELL.

It's very frequently mentioned in connection with SCAMs (Supplementary, Complementary and Alternative Medicines). Here you see mentions of "Quantum Therapy", "Quantum DNA", "Quantum Entanglement in Psychic Abilities" and the reliable old standby, "Quantum Healing".

When scientists do experiments with "Quantum Mechanics", they deal with very tiny processes that happen down at the scale of atoms and molecules – and usually in a vacuum and at ultra-low temperatures. So it's hard to take an ad seriously when it tries to sell a re-branded fridge magnet with the claim that it uses Quantum Magnetic Effects to heal everything from arthritis to varicose veins.

On one hand, health claims involving the word "Quantum" are rubbish. But amazingly, it seems that Quantum Mechanics doesn't just affect itty-bitty particles. It can ripple upwards to have effects on big living things like trees.

QUANTUM MECHANICS 101

But let's dive into Quantum Land, at the atomic scale, to see what it's really like. It's very different from where we live.

First, stuff is "quantised" – in other words, it can exist only in packets of fixed sizes. By way of comparison, in our world, a sound can have virtually an infinite number of loudness levels, and the same goes for the brightness of a light, and so on. But in Quantum Land, physical quantities (such as light, magnetism or location) can change only by fixed and discrete amounts. You can have one photon, or you can have two photons – but you can't have one and a half photons.

The actual word, "quantum", comes from Latin, and it means "how big" or "how much". This is a nod to the fact that "stuff" exists in discrete packets.

CRAZY QUANTUM

So many things about Quantum Mechanics are counter-intuitive (polite word for "crazy"). In that case, why do we bother with it? Unfortunately, because it works . . .

First, Quantum Mechanics successfully explains many details of the Universe around us.

Quantum Mechanics is the only tool that can explain the behaviour of that zoo of sub-atomic particles that makes up atoms and energy. (This zoo includes protons and neutrons and electrons, as well as photons.) We need it to understand how separate atoms combine to make molecules. For example, we can't properly understand covalent and ionic bonding unless we use Quantum Mechanics – except now we call it Quantum Chemistry.

Second, Quantum Mechanics is essential for our modern society to function. The laser, the memory stick, the simple transistor and the complex microchip, your smartphone, even the dumb light switch on the wall – they all depend on Quantum Mechanics.

HUMILITY BY NOBEL PRIZE WINNERS

Neils Bohr, one of the Gods of Quantum Mechanics, said, "If you are not astonished by Quantum Mechanics, then you have not understood it!"

Richard Feynman said, "I think I can safely say that nobody understands Quantum Mechanics."

Second, there is the strange concept of "Wave–Particle Duality". You might have heard that energy, depending on the circumstances, can be either a wave or a particle. For example, light can travel as a wave, but will interact like a particle. But did you know that *matter*, again depending on the circumstances, can also switch between being either a wave or a particle? (I am still astonished by this.)

QUANTUM AND CONSCIOUSNESS

In the early 1990s, Roger Penrose from the University of Oxford and Stuart Hameroff from the University of Arizona made a rather cheeky proposal. They asked, "Do our brains act as Quantum Computers? Is consciousness the result of the collapse of Quantum Superpositions?"

Is this possible? Yes.

Do we have any proof? Absolutely none, so far . . .

Third, the famous Heisenberg Uncertainty Principle. Up here, in our Big Non-Quantum World, a motorist can be busted for speeding. The Police Officer can be quite definite when they say that the car was at a

Steven Weinberg wrote, "There is now in my opinion no entirely satisfactory interpretation of Quantum Mechanics."

Mind you, these are quotes about how to *interpret* and understand Quantum Mechanics. The mathematics of Quantum Mechanics is extraordinarily precise, and it produces accurate predictions.

certain location, and travelling at a certain speed. This is not the case down in the atomic Quantum World. You can know quite precisely either a particle's Location, or its Momentum – but you can never know both at the same time. The same Uncertainty holds for Energy and Time.

Fourth, Measurement. In our Big World, you can measure the speed of a car, or the length of a kitchen table – and you don't change either the speed of the car, or the length of the table. But in Quantum Land, the mere act of measuring changes what you just measured. It's now different from what it was before you measured it.

Fifth, Quantum Entanglement. This confusing phenomenon happens when, say, a solitary sub-atomic particle decays into an "entangled" pair of smaller particles.

At the beginning, the original single particle had zero spin. To keep things balanced, the two daughter particles have opposite spins, which add up to zero. So you can't consider one particle without considering the other particle – they are "entangled". (I hope this makes sense so far.)

While you know that the spins of the particles have to be different, you don't know which particle has which spin. In fact, neither of them has any spin in a particular direction – until you actually measure them. (Whoa, now this is beginning to get weird.)

So let's say the two daughter particles get a huge distance apart. Measure one of them to find its spin. The act of measuring one particle forces the other particle – no matter how far away it is – to collapse into spinning in the other direction. And this happens instantly – with no delay. It happens faster than the speed of light. This is so weird that even Einstein was worried by it. He famously called it "spooky action at a distance".

JOHN STEWART BELL – DEEP PERCEPTION

Quantum Mechanics is confusing. Often physicists carry out an experiment, and then they argue about how to interpret it. (Some physicists say it's all about the interpretation.)

The physicist John Stewart Bell wrote a neat summary. "The problem is this: Quantum Mechanics is fundamentally about 'observations'. It necessarily divides the world into two parts, a part which is observed and a part which does the observing. The results depend in detail on just how this division is made, but no definite prescription for it is given."

He said that Classical Mechanics was quite different. "In Classical Mechanics we have a model of a theory which is not intrinsically inexact, for it neither needs nor is embarrassed by an observer."

THE BIG "BIG ONE" – LIFE

Back in the 1940s, scientists began to speculate that Quantum Mechanics might provide a link or a bridge between Non-Life and Life. Erwin Schrödinger, one of the Gods of Quantum Mechanics, published *What is Life?* in 1944. Other Quantum Heavies such as Neils Bohr, Werner Heisenberg and Eugene Wigner also pondered this idea.

But to quote the physicist and astrobiologist Paul Davies, writing in 2009, "Life seems little short of miraculous – all those stupid atoms getting together to perform such clever tricks . . . Today, we know that no special "life force" is at work in biology; there is just ordinary matter doing extraordinary things, all the while obeying the familiar laws of physics."

He goes on to say that, at this stage, we can't prove (or disprove) that Quantum Mechanics was part of that big jump from Non-Life to Life.

But we are pretty definite that Quantum Mechanics is involved with making some biological process work as well as they do. (In the meantime, old-fashioned ball-and-stick Chemistry and Physics explain most processes and phenomena in biology.)

PHOTOSYNTHESIS AND QUANTUM: QUESTION

Quantum is hard to believe. However, in Physics, Personal Belief and Opinion have no sway over the mathematical predictions.

And regardless of human beliefs, thanks to several billion years of evolution, Biology uses whatever works.

Consider "photosynthesis".

The word literally means using light (that's "photo") to make something (that's "synthesis"). So photosynthesis usually exploits the energy in light to grab carbon dioxide from the atmosphere, and then proceeds along a series of chemical reactions that end up making carbohydrates and oxygen. (Yep, carbohydrates and oxygen are essential if you want to eat and breathe.)

The whole process of photosynthesis is very complex, and we still don't fully understand it. But we know that the process begins with collections of atoms, which are formed into tiny "antennas". (To be technical, the proper name for an individual "antenna complex of pigment-protein" is the Fenna-Matthews-Olson (FMO) Complex, but I'll just call it an antenna.) These antennas catch the energy in the sunlight, and then very efficiently funnel this Excitation Energy through a huge forest of chlorophyll molecules to a Reaction Centre – where the next set of chemical reactions happen. Ultimately, the

THREE TIMES THE POWER OF THE HUMAN RACE

At any given moment, the process of photosynthesis captures about 130 terawatts of power from the Sun. That is roughly three times as much power as we humans generate and use. Over the period of a year, photosynthesis uses this power to grab about 100 billion tonnes of carbon and turn it into biomass and oxygen.

process finishes with the production of carbohydrates and oxygen.

But there's still one Really Big Mystery about photosynthesis. Why is it so fiendishly efficient? (I use the word "efficient" in its Engineering sense, as in, how much you get back out for what you put in.)

Most biological and human processes run at somewhat less than 50 per cent efficiency. With your car, only about one quarter of the energy embedded in the fuel ends up actually pushing the car down the road – the rest is wasted as heat, friction, and so on. But in the early stage of actually capturing the light energy, photosynthesis runs at around 95 per cent efficiency – amazingly high.

PHOTOSYNTHESIS AND QUANTUM – ANSWER

Let me emphasise that this research is still cutting-edge, and a little controversial.

As I said earlier, in the earliest stages of photosynthesis, the molecules gathering up the incoming light are arranged like antennas. They are arranged next to each other. When a photon of light energy lands on one of these antennas, it seems to stop being a single particle and instead turns into a smeared and spread-out wave.

$$6\,CO_2 + 6\,H_2O \xrightarrow{\text{light}} C_6H_{12}O_6 + 6\,O_2$$

Carbon dioxide water sugar oxygen

The critters that do photosynthesis are plants (both in the oceans and on land), as well as algae and cyanobacteria. These guys work hard for the money . . .

As a single particle, the incoming light "sees" only one antenna. But as a wave, it seems to overlap a whole bunch of antennas. This wave then somehow manages to sample all the possible different energy pathways, all at the same time – and nearly always pick the most efficient trajectory. (It's a bit like searching a database with a Quantum Computer. Instead of searching just one entry at a time, you search all of them at the same time.)

Now because this is all happening at warm temperatures, the atoms and molecules will jiggle around a lot. So over time, the most efficient pathway to carry the light energy will change – sometimes this pathway, and then, several tenths of a billionth of a second later, a different pathway. Even so, the light energy almost always chooses the most efficient pathway.

In 2013, scientists examined photosynthesis and managed to measure the fleeting existence of unexpectedly long-lived Quantum Coherence States. ("Coherence" means that the wave patterns of every part of a system stays in step together.) These Coherence States survived for about four tenths of a trillionth of a second. This is long enough for Quantum Effects to happen. This is more "evidence" that photosynthesis could happen via Quantum Effects.

QUANTUM AND BIOLOGY

Quantum Effects seem to be essential to the ability of some birds - specifically, the European Robin - to navigate on their long journeys by allowing them to sense the local magnetic field. These effects also seem to be involved in making the senses of hearing and vision in some other creatures as sensitive as they are. For example, when we humans

QUANTUM AND OTHER BIOLOGY

It also seems that Quantum Effects are necessary to life, not just for the energy collection process in photosynthesis, but also in enzymes.

There are many thousands of different enzymes in your body. They're essential for life. Enzymes are chemicals that accelerate chemical reactions by factors of trillions of times, or even more. It's long been a mystery as to how they speed reactions by so much.

Enzymes appear to use Quantum Tunnelling. Prior to this, we've seen Quantum Tunnelling in non-biological activity, such as radioactive decay, and also in how our Sun burns hydrogen to make helium and keep us warm. (Quantum Tunnelling is well-accepted and non-controversial. However, "Coherence" and "Entanglement" are still somewhat controversial.)

But it seems that an enzyme can encourage protons and electrons to actually vanish from one location on the enzyme's surface. And then instantly, with zero delay, they rematerialise (or "tunnel" through to arrive) in a different location on the enzyme. It's a kind of Quantum Teleportation.

Just as an aside, it seems that Quantum Tunnelling might be involved in the hydrogen bonds in our very DNA. Is this a hint that Quantum Effects are fundamental to our most basic chemicals of Life – and Life itself? Maybe.

listen to the quietest sounds we can detect, our ear drums move back and forth a distance equal to the diameter of a hydrogen atom – an almost impossibly small distance.

INEFFICIENT EVOLUTION

In photosynthesis, the early stages of gathering then funnelling the light to where it can be used are very efficient.

But the overall process of photosynthesis is not very efficient. For example, plants could catch more light if they were black. Instead, most of them are green – which means that they reflect the blue and red light.

Yes, the energy in the blue and red light is wasted. But that's evolution for you – usually settling for a Pretty Good Solution, not necessarily the Best Solution.

SCHRÖDINGER'S CAT

This Thought Experiment is an oldie, but it's a goldie.

Imagine that you have a sealed metal box. You have no way of knowing what is happening inside the box.

Inside you have a cat and a poisonous gas capsule that will release its contents at some random time. You have no idea when that time will be.

So what's the situation with our innocent cat?

In Classical Mechanics, the cat is either dead or alive – depending on whether the poisonous gas capsule has already burst open.

But in Quantum Mechanics, the cat is described by a wave function. This wave function describes the Quantum State of an isolated system. (It contains all the information about the system.) As a result, the cat is simultaneously both alive and dead at the same time – until you observe or measure it. When you observe the cat, its wave function then "collapses", and the cat is suddenly either dead or alive.

But, just to be contrary, perhaps the wave function is not real. It merely represents a State of Knowledge. The "collapse" is just what happens when you gain more information. (Not everybody agrees with this interpretation.)

With Schrödinger's Cat, the act of making an observation changes the situation.

FREE
CAT
FOOD!

?

Quantum Life < 45

06

NAKED MOLE RATS DON'T GET CANCER?

RODENTS MAKE UP ABOUT 40 PER CENT OF ALL MAMMALS ON EARTH – THEY ARE REALLY GOOD AT SURVIVING. THEY THRIVE IN ALL KINDS OF EXTREME ENVIRONMENTS.

But the rodent called the Naked Mole Rat is extra special. It doesn't seem to age, it doesn't feel some types of pain, it lives in communities similar to those of bees and ants – and it never, ever gets cancer.

NON-NAKED NON-MOLE NON-RAT

The name "Naked Mole Rat" is wrong on three counts.

This animal is not totally "Naked". (Important point: ignore the Common Cultural Definition of "naked" meaning "no clothes", and use the Zoological Definition of "no fur or feathers".) Sure, it's mostly covered with bare skin. But on the other hand, it does have about 100 hairs. In comparison, we humans have about five million hairs (mostly fine and short) that cover all of our body except for the palms and feet.

Also, the Naked Mole Rat isn't a "Mole" – that's an entirely different mammal.

Finally, even though it *is* a rodent, it's not a "Rat". It's more closely related to porcupines and guinea pigs than to rats.

NAKED MOLE RAT 101

Unkind people have described them as "sabre-toothed sausages", but for me, Naked Mole Rats are so ugly that they fit into the "beautiful" category.

They have no sweat glands, and no subcutaneous fat. They never drink.

Naked Mole Rats live in the dry high plateaus of East Africa. They've been there for at least 2.5 million years. They're small – about 8–10 centimetres long, weighing about 35 grams on average. Breeding queens can be twice the mass of the males. Surprisingly, if a non-breeding adult comes near a pregnant female, his or her nipples will start to develop. So the breeding queen can exert powerful effects on others.

They live in groups of as few as 20 to as many as 300 individuals, but mostly around 80 or so. They dig underground tunnels that are kilometres long – just with their incisor teeth. In fact, about one quarter of all their muscles are in their jaw. (We humans carry one quarter of our muscles in our legs.)

Their eyes can't detect details. They see only the difference between light and dark. This is probably because they never have to leave the dark safety of their tunnels.

Naked Mole Rats are officially "vegetarians", mostly eating bulbs and tubers. However, they do eat their own dead babies (made from meat). They also eat their own faeces – and if they're not eating it, they're rolling around in it, hence their characteristic smell.

LONGEST MAMMALIAN TUNNELS

The Zambian Mole Rat holds the record for the longest non-human mammalian tunnel. These tunnels can reach 2.8 kilometres.

Zambian Mole Rats are also quite co-operative with each other. In one case, four colonies had linked up, creating 7.2 kilometres of tunnels in total.

However, there are only 10 or so Zambian Mole Rats in each colony. So some poor Mole Rats must have done a lot of digging…

LIKE INSECTS

The Mole Rat is the only mammal to be "eusocial". Eusocial behaviour is usually found only in insects, such as bees, termites, ants and wasps.

Naked Mole Rats Don't Get Cancer? < 49

"Eusocial" means that a single female, the queen, does all the breeding, and all the other members of the community keep the colony functioning. The Naked Mole Rat queen is serviced by up to three males.

If the breeding queen is removed, another female will arise and take her place. This behaviour also happens in a bee hive.

Again as in a bee hive, all the other Naked Mole Rats living in the burrow are organised in rigid castes. Some dig tunnels, some are warriors, some look after the babies, some forage for food, and so on.

The other two eusocial mammals (that we know of) are relatives: the Zambian Mole Rat (which can have a few queens in each colony), and the Damaraland Mole Rat.

PAIN?

Most mammals feel pain from acid.

Like us, Naked Mole Rats will feel the pain from a cut or a burn. But unlike us, they don't feel chemical pain, such as from acid, capsaicin (the stuff that makes chilli hot), or lemon juice in a wound.

This is probably an adaptation to their living environment. Their underground tunnels aren't well ventilated. As a result, the carbon dioxide level is not the 400 parts per million (0.04 per cent) we surface-dwellers currently enjoy, but 50,000 parts per million, or five per cent. (We humans can function for only about four hours in levels that high, but Naked Mole Rats do it all their lives.)

The chemical problem is that carbon dioxide dissolves in water to make carbonic acid. The more carbon dioxide, the stronger the acid. Not only does carbonic acid disrupt normal physiological functioning, it's also painful.

So the Naked Mole Rat has evolved to not feel any chemical pain under these conditions of high carbon dioxide. Thanks to a special adaptation in their pain sensors – sodium channels, if you want to look it up – they can function normally in an acidic environment for their whole lives.

There is a potential benefit in knowing this.

In humans, the nerves that carry chemical pain are linked to post-traumatic pain. This type of pain includes joint pain after a knee injury, or surgical pain. If we get better at stopping chemical pain, we might be able to stop post-traumatic pain as well.

METHUSELAH

Speaking of lives, Naked Mole Rats are incredibly long-lived. They can live for over 30 years – seven times longer than expected for an animal of their size.

Surprisingly, during that long life, they don't show any signs of aging. They don't exhibit osteoporosis, muscular fraility, mental decline, changes in basal metabolic rate or body composition, heart disease, menopause or even slowing down of various organ systems. Breeding queens deliver pups until they die.

There are many teams of scientists trying to solve the puzzle of their prolonged youthfulness.

There's one line of research that shows their DNA and RNA are different from that in other animals. As a result, proteins are made with extra fidelity and accuracy – not just in their youth, but for their whole lives. Perhaps the presence of fewer aberrant proteins contributes to their remarkable longevity?

Regardless of the mechanism, Naked Mole Rats stay strong and robust until The End.

NO CANCER

And speaking of The End, it's not going to be caused by cancer.

Cancers account for about 10–15 per cent of all human deaths, and about 90 per cent of all mice and rat deaths (providing they can avoid befriending cats and the like).

Naked Mole Rats? Zero cancer deaths. They simply don't get any cancers.

MECHANISMS OF NO CANCER

Cancer cells can "win" over regular cells simply because they never stop replicating or growing.

Under normal circumstances, the cells in your body are always dying and always regenerating. Usually, they regenerate and grow to a certain stage. Once they have replaced what was there originally, they stop growing.

However, cancer cells do not obey this rule. In every known animal cancer cells never stop growing – except inside the Naked Mole Rat.

One pathway Naked Mole Rats might use to avoid cancer is Early Contact Inhibition. "Contact" refers to cells touching each other. In the Naked Mole Rat, when cells get too close to each other, they stop growing.

If you get a bunch of normal human cells and grow them on a culture plate, they'll grow until they form a single thin monolayer on that plate – and then they'll stop.

But cells from the Naked Mole Rat won't even get to the single thin monolayer stage. Once they sense another Naked Mole Rat cell is nearby, they'll stop growing. So if a cell inside a Naked Mole Rat turns cancerous, it cannot keep on growing without limit. Any cancerous cell just "withers on the vine".

Another possible pathway of cancer resistance relates to their skin. Naked Mole Rats need a tough, elastic and flexible skin for their subterranean lifestyle. Their skin has these necessary characteristics because it contains a supersized complex sugar called HMM-HA (or High-Molecular-Mass Hyaluronan). It seems that HMM-HA stops cancer cells dead in their tracks.

These tiny, blind and bald Methuselahs, Naked Mole Rats, have so much to teach us about living longer and more robustly, how to bypass certain types of pain, and how to foil cancers – and that's a bare fact ...

COLD BLOODED

On average, the body temperature of a Naked Mole Rat hovers around 32°C.

But here's the surprise – they are not warm-blooded, like all the other mammals.

All other mammals use energy to maintain a set temperature (thermoregulators). For example, we humans shiver when cold, and sweat when hot.

Not so for Naked Mole Rats. They are "cold-blooded", like snakes (thermoconformers). If it's cold, they huddle together for warmth. If it's hot they move into the cooler depths of their tunnels.

But most of the time, they don't need to worry. From a hot Summer day to a cold Winter night, the temperature inside their tunnels stays within a narrow 2–3 degree range.

Naked Mole Rats Don't Get Cancer?   < 53

07 STINKY SYNTHETIC SHIRTS

I LOVE WEARING BRIGHT SHIRTS. (WHY SHOULD WOMEN HAVE ALL THE FUN?) BUT EARLY ON, I LEARNT AN EMBARRASSING LESSON – POLYESTER CLOTHES STINK AFTER SWEATING, NO MATTER HOW COLOURFUL THE PRINT.

It turns out that fresh sweat has little odour. (The long-chain fatty acids in sweat are too large to trigger your Olfactory Epithelium – the part of your nose that turns incoming small chemicals into electrical signals that go to your brain.)

A bacterium called *Corynebacterium* causes Smelly Armpit – on your skin. But while *Corynebacterium* is plentiful in the armpit, it won't grow well on textiles.

Then there are *Staphylocci* bacteria that live happily on both armpit skin and textiles. But they're not very stinky – they tend to generate only a "normal" non-stinky Body Odour.

Bacteria called *Micrococci* are the real Stink-Masters. They can break down the long-chain fatty acids, the hormones and the amino acids that are naturally present in sweat. In doing that, they create smaller and more volatile chemicals – which generate the typical Offensive Body Odour smell.

Micrococci grow really well on polyester (dunno why, yet). They don't grow well on cotton (dunno why, yet).

SO WHAT ARE THE SOLUTIONS TO SMELLY ARMPITS?

1. Wear cotton instead of polyester – and deal with sometimes wearing colours and patterns that are less vibrant.

2. Wear polyester if you love it, but add deodorant. And no, deodorants do not cause cancer. (See "Anti-Perspirant and Cancer" in my 23rd book, *Great Mythconceptions*.) But deodorants can encourage *Corynebacteria* to grow more in your armpits, leading to smelly armpit skin.

3. Dr Chris Callewaert of Ghent University is exploring a different solution – he wants to transplant bacteria from non-stinky people to their stinky relatives.

Remember – polyester pongs stronger, but you can wear cotton longer.

SKIN BACTERIA

The skin "microbiome" is very varied, with many niches, each with its own specific bacteria. Very dry locations such as arms and legs might carry only 100 bacteria per square centimetre. But more moist areas, like your armpits, belly button and toe web spaces, might carry 10 million bacteria per square centimetre.

Stinky Synthetic Shirts < 55

BLUE DRESS ILLUSION

IT'S VERY RARE INDEED WHEN VISUAL NEUROSCIENCE AND TEXTILE TECHNOLOGY COMBINE TO TAKE OVER THE INTERWEBS – BUT IT ALMOST HAPPENED IN FEBRUARY 2015. YUP, I'M TALKING ABOUT THE FAMOUS BLUE DRESS (OR WAS IT WHITE?).

The Blue Dress that almost brought the Web down is the one that "changes" colour all by itself, from blue to white. (By the way, in real life, the Blue Dress resolutely stays blue. The blue–white change happens only in that specific photo. This gives us a hint that there's a genuine visual illusion going on.)

At first, I though it was just a big beat-up. But when my friends saw a white dress while I saw a blue dress on the same computer screen, I realised that this was a lovely, and brand-new, Optical Illusion. It cried out for an explanation.

So, here's the short answer.

If you perceive the dress to be in shadow and back-lit, you'll "see" it as white. But if you believe the dress to be lit from the front, you'll "see" it as blue.

BLUE DRESS HISTORY

It all began with a wedding on the Scottish Isle of Colonsay, between Grace Johnston (née MacPhee) aged 27, and Keir Johnston, aged 29.

The mother of the bride, Cecilia Bleasdale, had bought a royal blue "Lace Bodycon Dress" from a British retailer, Roman Originals, for £50. About a week before the wedding, she sent the now-famous photo of the dress she planned to wear at the wedding to her daughter, Grace. She and her fiancé, Keir, had perhaps their first ever disagreement – about the colour of the dress. Was it blue, or white?

So Grace posted a photo of the dress to her Facebook page – for back-up. Their friends had the same colour problem – some saw white, others saw blue. But the upcoming wedding was more important, and interest in this strange phenomenon faded into the background.

Enter Caitlin McNeill, a 21-year-old singer and guitarist with the Scottish folk band Conach, which played at the wedding. She was a close friend of both the bride and groom. Caitlin and the rest of the band were fascinated by the photo. The dress was "obviously blue and black" when they looked at the actual dress – but the photo was not as straightforward. Caitlin said that they "almost didn't make it on stage because [they] were so caught up discussing this dress".

On 26 February 2015, Caitlin posted the soon-to-be-famous photo to her Tumblr blog. By that evening, this Optical Illusion had become famous. At its peak, over two thirds of a million people were looking at this photo at the same time on BuzzFeed. BuzzFeed realised this could generate massive ad revenue, and assigned two editorial teams to write more about the Blue Dress. By 1 March, BuzzFeed had received over 37 million hits about the Blue Dress. It was mentioned in over 10 million Tweets.

Celebrities were polarised in their views. Taylor Swift saw a blue dress, and said she was "confused and scared" by the whole business. Kim Kardashian saw a white dress, while her husband Kanye West saw a blue one.

So what's going on? It's a five-part answer – with four definite Do-Knows, and one Don't-Know.

PART 1: DRESS IS BLUE

First, the dress is actually blue. It looks blue in real life. And if you use the image-editing program Photoshop, you can analyse individual pixels of the photo and see that the Blue Dress is in fact blue. Indeed, the retailer, Roman Originals, advertised the dress as being blue.

On the other hand, about 68 per cent of BuzzFeed users saw the Blue Dress as white.

PART 2: COLOUR CONSTANCY

In chess boards, you expect the squares to be alternating light and dark. In the Checkerboard Shadow Illusion, created by Edward H. Adelson, John and Dorothy Wilson Professor of Vision Science at MIT, you can "see" that the vertical shape is casting a shadow over the square labelled "B". This influences what you "see". And yes, B looks a different shade from A.

But they are exactly the same shade. Part of this "fooling" happens because the eye–brain combination tries really hard to maintain what the visual neuroscientists call "Colour Constancy".

Consider a white sheet of paper. It just reflects whatever the ambient light colour is. It will be white in bright sunlight, but under the red lights of a nightclub it will actually reflect red light. It will look red.

But this change of colour bothers your brain. So your brain has evolved Colour Constancy, where it "adjusts" or compensates for the ambient light. In a nightclub, colour constancy removes the reddish influence of the nightclub lights, and suddenly a sheet of paper looks white – even though it's reflecting red light and actually looks red. Your eye–brain combination is ignoring reality, and giving you what you expect to see. (The Professional Photographers and Photoshop Experts call this "Shifting the White Balance".)

Colour Constancy is a survival advantage. It makes a red apple always look the same colour, whether that particular food item is in the shade or in sunlight. So Colour Constancy makes it easier to find food.

But sometimes Colour Constancy can be fooled – especially if what you're looking at gives you confusing signals. (This is a hallmark of a good Optical Illusion. For another weird and wonderful Optical Illusion, see the endpapers on the front and back of this book.)

Blue Dress Illusion < 61

PART 3: CONFUSING PHOTO

The Blue Dress photo tells you two contradictory "facts" at the same time.

It tells you that the Dress is lit from the front, but at the same time, the photo tells you the Dress is lit from the back.

How?

Purely by accident, the photo of the Blue Dress is beautifully ambiguous. There is absolutely no bare skin – which always gives you a good reference as to the true colour. There are no other dresses, such as a white wedding gown, or other people, that could give you a clue.

All you see is the fabric of the blue dress – and an out-of-focus band of background brightness on the right side of the photo. This might make you think that the front of the blue dress is in shadow.

But the bolero jacket at the top of the dress contains a panel of shiny fabric that is partly reflective. I'm pretty sure that this is essential for helping to create this Optical Illusion. Visual neuroscientists call these mirror-like reflections on the shiny part of an object "specularities". It turns out that specularities can give you the best clue as to the actual colour of the ambient light. And in this case, the specularities give you the impression that the dress is well illuminated from the front.

PART 4: BIG FINISH (ALMOST)

We can now put it all together.

If you assume that the front of the dress is in shadow (thanks to the bright, blurry background light), your brain will apply Colour Constancy to remove the bluish hue of the shadow – and bingo, the dress is white.

But if you assume that the dress is front lit (thanks to the shiny reflections on the top panel of the bolero jacket), you will see the dress as blue.

So that's what we know.

PART 5: WHAT WE DON'T KNOW

And now for what we don't know (because it's good to have some humility).

Why do some people assume shadow and a white dress, while others assume brightness and a blue dress? At this stage, we simply don't know.

As far as we do know, it's not related to your current state of mind, or your emotional or intellectual intelligence.

But don't forget, as in all good Magic, there is a slight misdirection. Many people assumed that the dress belonged to the bride. In Western society, the bride usually wears white. If it had been made clear that the dress belonged to the bride's mother, we might have been influenced by the cultural assumption that she would be very unlikely to choose to wear white.

The Talmud writes, "We do not see things as they are; we see things as we are." As in all visual illusions, we've been blinded by the light...

09

BRONTO-SAURUS IS BACK

MOST OF US HAVE A BIG SOFT SPOT FOR DINOSAURS, AND HAVE PICKED UP A BUNCH OF FACTS ABOUT THEM. AMATEUR DINOSAUR ENTHUSIASTS – ESPECIALLY EIGHT-YEAR-OLDS – KNOW THAT THE NAME *BRONTOSAURUS* IS JUST PLAIN WRONG.

The correct name, apparently, is *Apatosaurus*. Well, it doesn't happen often, but it seems that the eight-year-olds are wrong, this once.

DINOSAURS 101

Dinosaurs came in three main body shapes.

There was the two-legged dinosaur, the four-legged dinosaur with a short neck, and the four-legged dinosaur with a long neck (sauropod).

Both *Apatosaurus* and *Brontosaurus* belonged to this last category – they had four sturdy legs, really long necks and long tapering tails, and small heads. (Many dinosaurs beside *Brontosaurus* and *Apatosaurus* fitted into this sauropod category – such as *Brachiosaurus*, *Diplodocus* and *Camarasaurus*.) The charismatic *Brontosaurus* was a little smaller than *Apatosaurus*.

They were vegetarians who lived about 160–145 million years ago. They came in different sizes, but the really big ones were some 35 metres long and weighed over 40 tonnes. They were amongst the largest land animals ever to walk the Earth.

BONE WARS: PART 1

These dinosaurs were discovered as a result of the infamous "Bone Wars", also called the "Great Dinosaur Rush".

Over a 20-year period from around 1872 to 1892, two American paleontologists, Edward Drinker Cope and Othniel Charles Marsh, engaged in a bitter competition (using various underhanded methods) to discover and describe more dinosaur fossils than the other.

Before they started, there were only nine named species of dinosaur in the USA. By the end of the Bone Wars, between them, Marsh and Cope had discovered and described some 142 "new" species of dinosaur. However, only 32 or so are considered valid species today – it seems their bitter rivalry led to shoddy science.

Marsh and Cope first met in Berlin around 1864. Initially, they were quite friendly, spent time together, and even named new species after each other. But over time, their differences contributed to what became an acrimonious affair.

Othniel Charles Marsh supported Charles Darwin's Theory of Evolution, and tended towards being methodical, and somewhat introverted. He had grown up in a poor family. But he was lucky enough to have a wealthy uncle, the businessman and philanthropist George Peabody. (In fact, Peabody is known as the Father of Modern Philanthropy.) Peabody built the Peabody Museum of Natural History at Yale University, made his nephew Marsh the head of the museum, and even left him an inheritance after he died. This position and the later inheritance left Marsh financially secure.

On the other hand, Edward Drinker Cope came from a wealthy and influential family, supported Neo-Lamarckism (a theory about inherited characteristics) rather than Darwin's Theory, and tended towards having a quick temper. Despite having little formal education, Cope became a professor at the Academy of Natural Sciences in Philadelphia.

BONE WARS: PART 2

It's said that there's a fine line between love and hate.

Cope and Marsh's competitive desire to be the first with a new dinosaur discovery meant that their initial cordiality soured into outright antagonism and resentment.

Their battlefield was the newly discovered fossil beds of the American West in Nebraska, Wyoming and Colorado. They would excavate in the summer, and publish in the winter. They each used their personal wealth to mount their own expeditions, but they also bought dinosaur fossils from third parties. Soon they were sending tonnes of dinosaur fossils back to their bases on the East Coast of the USA.

They used bribery and theft to access dinosaur fossils, stole from each other, and even destroyed bones and covered up digging sites.

HARSH, MARSH

Perhaps it was because Professor Marsh had so much practice digging up dirt looking for bones, that he wasn't afraid to dish the dirt on Professor Cope?

One comment accuses his arch-nemesis, Cope, of taking Marsh's ideas and passing them off as his own.

"Prof. Cope has recently claimed this discovery on the strength of a paper which he read before the Spring Meeting of the National Academy, in 1876. He knew, however, at the time that my article was already published, and during the reading of his paper, a printed copy of my publication was in the room, in the hands of a member."

On occasion, their rival digging teams would literally cast stones at each other. They each attacked their opponent in print, trying to ruin their rival's professional credibility by making claims of outright errors, plagiarism and financial mismanagement.

In their efforts to outdo each other and to gain scientific precedence, they would dash off hasty telegrams briefly describing their finds, and only later formally publish a fuller account. As a result, their science was often haphazard and careless. For example, they would sometimes almost-randomly assemble the bones of two – or even more – different species into the same skeleton. On occasion, they would even separately describe the same dinosaur, giving it different names.

In their later years, they were each financially ruined by their unrelenting rivalry and shameful conduct. In fact, Cope even tried to continue the battle with Marsh from beyond the grave. He specified that after he died, his brain was to be measured to show that his brain was larger than Marsh's brain – and therefore, Cope was supposedly more intelligent. Marsh, however, never took up the challenge.

Another comment accuses Cope of amateur-level mistakes in identifying characteristics of bones.

"Prof. Cope, mistaking the character of these vertebrae in an allied form, described them as representing a new genus, *Amphicoelias*, and even a new family, *Amphicoeliidae* . . . All the known *Sauropoda*, however, have similar vertebrae, with opisthocoelian centra in the cervical and anterior dorsal regions."

BRONTOSAURUS EMERGES

So let's get back to the *Brontosaurus* versus *Apatosaurus* stoush. Marsh named them both.

In the December 1877 issue of the *American Journal of Science*, Marsh described and named three new dinosaurs: *Stegosaurus*, *Allosaurus*, and yes, *Apatosaurus*. *Apatosaurus* means "deceptive lizard".

Two years later, Marsh published a brief note about another dinosaur which was somewhat similar, and which he called *Brontosaurus*, meaning "thunder lizard".

LUMPERS AND SPLITTERS

The basic problem for *Brontosaurus* was that it was quite similar to *Apatosaurus*. So after a bit of discussion, the name *Brontosaurus* got sent into the shadows – for over a century.

Now, the field of classifying living (or extinct) creatures or plants is called "Taxonomy". In Taxonomy, there are "Lumpers" – and their exact opposite, "Splitters". The Lumpers like to lump life forms together to end up with just a few groups. But the Splitters like to find differences between life forms, and create lots of groups.

WHAT'S IN A NAME?

In 1903, about a quarter of a century after Marsh had classified *Apatosaurus* and *Brontosaurus* as two separate genera (the plural of "genus"), the Lumpers in the paleontological community won.

In that year, another paleontologist, Elmer Riggs, claimed in the *Geological Series of the Field Columbian Museum* journal that there were not major differences between these two dinosaurs after all. As a result, both *Apatosaurus* and *Brontosaurus* were placed into the same genus, *Apatosaurus*.

You see, there are rules in the field of Taxonomy (thanks to the 18th century Swedish systematicist, Carl Linnaeus). One of the rules is that if there is a conflict, the name that was given first gets priority. So in 1903, *Apatosaurus* won and *Brontosaurus*, even though it was a really cute name, lost.

However, because it was a really cute name, the general public kept on using it. However, the paleontologists were steadfast. In fact, in 1978, further detailed research by John S. McIntosh and David Berman confirmed that the name *Brontosaurus* should be deleted and go the way of the dinosaurs. (However, according to my 29th book, *Dinosaurs Aren't Dead*, dinosaurs didn't die out. Modern birds are the direct descendants of a line of dinosaurs that didn't die out. And yes, *Velociraptor* had feathers).

US POSTAL SERVICE STUFFS UP

But nobody told United States Postal Service.

About a decade later in 1989, they released a stamp with the word "*Brontosaurus*" proudly emblazoned upon it. The fact that the stamp was officially launched at Disney World (definitely not a "proper" museum) didn't help them with getting a Scientific Stamp of Approval. There was a huge uproar.

Surprisingly, there was a much bigger "mistake" in this stamp release that the media didn't comment on. There were four stamps issued in this Dinosaur Series. One of the stamps was of the *Pteranodon*. But

Pteranodon is a flying reptile – it is not even a dinosaur!

Maybe the furore was a good measure of the interest and secret affection that people had for the name *Brontosaurus*.

THE HARDER YOU WORK, THE LUCKIER YOU GET . . .

Even so, the name *Brontosaurus* was buried again – until some Portuguese and English paleontologists decided to try to better understand the evolutionary relationships linking these various four-legged, long-necked, plant-eating dinosaurs.

Emanuel Tschopp had chosen this topic for his PhD thesis at the New University of Lisbon. He was trying to generate a high-resolution family tree of the sauropods.

He and his colleagues were extraordinarily thorough.

They measured some 477 individual features of fossilised dinosaur bones – features such as length, bumps, distances between these bumps, shapes of these bumps, and so on. Now, there are a lot of bones in a dinosaur. So they did this for every bone of each of 81 separate relevant sauropod skeletons stored in some 20-plus museums scattered around the world. The amount of work to be done was simply enormous. While some of the bones are huge and heavy, they are also fragile, so collecting data can be painfully slow. This monumental task took five years.

At the end of it, the bones "spoke" to them – and gave up their secrets.

The paleontologists now (more than ever before) had a deeper understanding of how to classify these long-necked herbivores. And getting back to *Brontosaurus*, this led to a slight "re-arrangement" of the family tree.

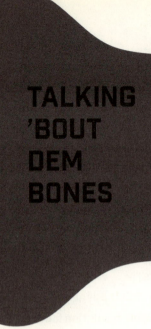

TALKING 'BOUT DEM BONES

While paleontology is a lot of fun – hey, it deals with dinosaurs – you might think that it's not very useful to most of us, apart from helping find new mineral deposits. But fossils tell us stories – such as the story of life.

BACK FROM THE SHADOWS

Dr Tschopp, the lead author of that paper, later said, "I didn't start out trying to resurrect *Brontosaurus*." But accidentally, that's what he did.

He had to. There were so many significant differences in the bones that *Brontosaurus* was split off again from *Apatosaurus* into its own separate genus. For example, *Apatosaurus* had a more robust body and a more bulky neck than *Brontosaurus*. *Brontosaurus*, which was now split even further into three separate species, had longer bones in the ankle compared to *Apatosaurus*, and had a rounded edge on the shoulder blade.

Over a period of more than a century, *Brontosaurus* rose, fell, and rose again. But now, thanks to five years of very detailed research, and a gargantuan scientific paper nearly 300 pages long, the Splitters have won – and the Lumpers just have to lump it. *Brontosaurus* is back out of the shadows.

But unfortunately, they have resuscitated only the name, not the actual living dinosaur...

They tell us that some 250 million years ago, there was a truly enormous release of methane into the atmosphere from parts of Siberia. Methane is a powerful greenhouse gas – much more powerful than carbon dioxide. So much methane was released that atmospheric temperatures rose by 16°C. Life went through one of the greatest mass extinctions of all time.

TAIL = BULLWHIP

Polymath and ex-Chief Technology Officer at Microsoft Nathan Myhrvold has a very interesting theory. He reckons the *Brontosaurus* tail could have been used for self-defence.

He noted that the very long tail was a metre thick at its base, but towards the tip it tapered to merely the diameter of a garden hose. It was very similar to a whip. He wondered what would happen if the *Brontosaurus* shook its tail.

A certain amount of energy would be "injected" into the tail at its massive base – which would move. Very little energy would be lost, so practically all the energy would make its way to the slender tip. Because the tip weighed so little, it would move very fast (to keep the amount of energy constant).

Nathan calculated that the tip would move at around 1260 kilometres per hour – faster than the speed of sound! The sound level would have been around 200 decibels – much louder than a 747 jet taking off at full power right next to you. It would be as loud as a massive cannon.

He reckoned that it was possible that Brontosaurs could "crack their tails like bullwhips, creating sonic booms to scare away predators".

10

STARS ARE POINTY

WHETHER IT'S THE STAR WE PLONK ONTO THE CHRISTMAS TREE, THE STARS THAT CAN ADORN OUR FLAGS AND PYJAMAS, OR THE STARS ON THE HOLLYWOOD WALK OF FAME – THEY ALL HAVE POINTS!

But we know that a real star doesn't actually have any points or spikes. A star is a giant spherical ball of plasma. Furthermore, all the stars that we can see (apart from our Sun) are so far away that they appear to us as perfect little dots.

So why do we draw stars as pointy objects? It's because our eyes actually see them as having points. And why is that? Because the lens in each human eyeball has two imperfections called "Suture Lines".

EYE 101

A human eyeball is about the size of a golf ball – 25 millimetres or so.

At the very front is the cornea. It does about two thirds of the "bending" of incoming light. This light needs to be bent so that it lands exactly on the retina – not in front, nor behind.

The lens does the remaining one third of the bending of the light. The lens has an Optical Power of 18 dioptres. (To measure is to know). Human height and weight are familiarly measured in metres and kilograms. The "bending ability of a lens" is measured in the probably unfamiliar unit of "dioptre".

While the Optical Power of the cornea is fixed at around 43 dioptres, that of the lens is adjustable. So it's the lens that ensures that the image sits exactly on the retina.

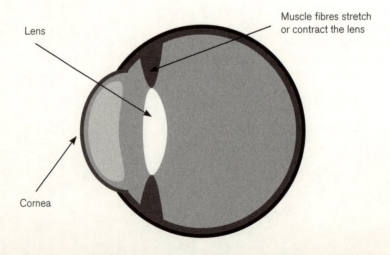

LENS 101

This lens inside your eyeball looks like an asymmetrical squashed sphere – it's more flattened at the front, more rounded at the rear.

Front-to-back, it's about four millimetres deep, but side-to-side, the lens is about 10 millimetres across. This widest part is called the Equator.

Thanks to little muscles attached near the Equator, the lens can change its shape. It can do this in one third of a second. So as you look at distant, and then near, objects, the lens changes shape to ensure that the image always lands exactly on the retina.

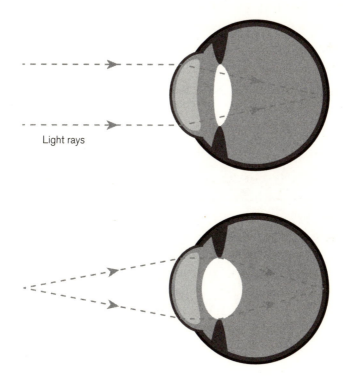

Light rays

But it's only when you are very young that your lens can change its shape enormously – and vary its Optical Power by up to 15 of its 18 dioptres.

By the time you reach 45–50 years of age, your range of adjustment (technically called "Accommodation") is down to only two dioptres. At this stage, if you want to look at both distant and near objects in focus, you need to wear glasses. By the time you reach 70, all you get is one lousy dioptre of Accommodation.

LENS 102

The lens is almost totally transparent. It has three main parts.

First, the lens capsule wraps around the lens, and holds it together. This membrane is a bit like a plastic bag. It's made from proteins (such as collagens) and sugars (chains of repeated disaccharides, called sulfated glycosaminoglycans.) The lens capsule is incredibly thin. It varies between two and 28 microns in thickness. (By comparison, a fibre of hair is about 50–70 microns thick).

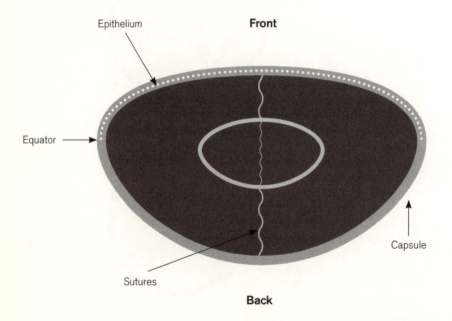

Second, the lens epithelium is a thin layer of simple cuboidal cells, located just under the capsule, and only at the front half of the lens. This layer of cells reaches back to the Equator of the lens. Besides other tasks, the lens epithelium regulates the volume and saltiness of the lens. At the Equator, its cells also make new lens fibres that migrate toward the centre-line of the lens. (If you want to learn more, look up "lens morphogenesis".)

And third, there are lens fibres that make up the bulk of the lens. They are actually transparent cells that are both long (up to 12 millimetres) and skinny (3 – 10 microns in diameter). These fibres meet and blend at Y-shaped Suture Lines. There are two of these Suture Lines. The front one is a regular Y-shape, while the back one is an upside-down Y-shape.

These Suture Lines bend the waves of light as they travel through, and past, them.

DIFFRACTION 101

This bending is called "diffraction". Whenever light passes around an object, or through a slit inside an object, it bends and changes its direction.

Now here's an example of pointy stars you would have seen dozens of times, and never noticed.

You've heard of the famous Hubble Space Telescope, which was launched in 1990. Whenever there's a star in a picture produced by this telescope, that star has four points or spikes – even though the incoming image is actually a round little dot.

Why the points (and always four of them)?

Because in the Hubble Space Telescope, the smaller secondary mirror is held in position by *four* cross hair-like struts, and the incoming light has to travel past these struts to land on the bigger main mirror. These struts bend the light, giving the star its characteristic four points.

A similar effect happens with your eyes, thanks to the Suture Lines inside the lens of your eyeball bending the incoming light. But, because they are "organic", every human eye has slightly different Suture Lines. Even your right and left eyes have different Suture Lines – so you might see four points with one eye, but five with the other.

But the point of it all is this – there's nothing like being dazzled by a little starlight . . .

MINI-RAINBOW POINTS

Nature has another surprising delight for us in the points around a star. The points are mini-rainbows!

It turns out that diffraction bends red light more than it bends blue light. (After all, in a rainbow, the red band is on the outside, while the blue band is on the inside.)

Look very carefully at a single point of a star. (This is a lot easier with a photograph than in real time. This is because a photograph can collect the light for a few seconds, or minutes, making the image brighter.) Each point is red on the outside, and blue towards the centre.

This is yet another case of the Economic Invisibility of Nature – it gives us wonderful stuff for free.

11 CORN PASSES STRAIGHT THROUGH

HOW COME, WHENEVER YOU EAT CORN KERNELS, THEY MAGICALLY REAPPEAR, TOTALLY INTACT, IN YOUR TOILET BOWL? AND THIS HAPPENS NO MATTER HOW CAREFULLY YOU CHEW THE DELICIOUS CORN KERNELS – RATHER THAN GULPING THEM DOWN.

The answer is that what you see in the toilet bowl is the indigestible and tough outer covering of the corn kernel – the pericarp.

Pericarp is made from individual monosaccharide sugar molecules stuck together – cellulose. But the way that these sugars are joined makes them immune to being broken down by the enzymes in your gut. A cow has a bunch of stomachs (equivalent to fermentation chambers) that can separate these individual monosaccharides. But we're not cows, so we can't do this.

Starch is similar to cellulose in that they are both made from individual monosaccharide sugar molecules stuck together. But the difference is in how the sugar molecules are stuck together. Our gut makes enzymes that can easily split the sugar molecules in starch away from each other. As a result, we get energy from starch. But we cannot get energy from cellulose (which is in the pericarp).

Which is why yellow bits pass all the way through your gut into the toilet bowl.

Mind you, while cellulose is indigestible, it's not useless. Pericarp is "fibre", which absorbs water in your gut. This means that it gives you a sense of fullness, so you stop feeling hungry. The absorbed water also makes your stools soft and bulky – and easier to pass.

According to US figures, you actually get pretty good nutrition from corn.

A cup of corn kernels will give you about 4–5 grams of protein, 1–2 grams of fat, about six grams of sugars, and about 25 grams of starch.

Overall, a cup of corn gives you about 600 kilojoules of energy – about seven to 10 per cent of your daily requirements (again, these are US figures).

So what comes through when you eat popcorn – pop or poop?

12

NAKED IN THE ANTARCTIC

THERE ARE SOME PRETTY EXCLUSIVE "CLUBS" AROUND. FOR EXAMPLE, ONLY A FEW HUNDRED PEOPLE HAVE EVER LEFT A SPACECRAFT TO ENTER THE COLD VACUUM OF SPACE FOR A WALK.

Another elite club is the 300 Club – and it's pretty cool. The entry requirements are brutally simple – and scary. You join this club only by exposing your skin to a sudden temperature change of slightly more than 300F°. That's over 166C°.

You don't join by warming up your skin, say with a blowtorch. No, you use the natural and extreme cold of the South Pole – in the dark of winter.

HISTORY OF 300 CLUB

The 300 Club is the most southerly club on the planet.

This traditional winter ritual happens down in the Antarctic, at the Amundsen-Scott South Pole Station. It's Do-It-Yourself entertainment for the lonely Antarctic residents, cut off from the rest of the world in the depths of winter. This Antarctic Base is the most remote base operated by humans – and by "most remote" I mean anywhere in the known Universe, including the International Space Station. Each winter, about 50 people live and work there. For seven out of every 12 months (mid-February to mid-September), there's no way in or out.

The 300 Club began as the 200 Club. Back in 1959, a few of the staff turned an empty wooden packing crate into a crude sauna. They ran it up to about +120°F (about +49.9°C). On a day colder than -90°F (-67.8°C), they would trot out of the hot sauna to circle around the South Pole. That exposed their skin to a temperature change of over 200F° (actually 210F°) – hence the name.

But technology has improved, and so now it's the 300 Club.

ENTRY TO 300 CLUB

The 300 Club has a number of prerequisites.

First, the Government Meteorologist has to proclaim that the outside air temperature is officially a bit colder than -100°F (about -73.3°C). Then, inside the Station, they crank up the sauna to a little hotter than +200°F (about +93.3°C). You get in the sauna, naked apart from insulated "bunny" boots, and sweat it out for a few minutes. Then, you towel away the sweat and head outside for a bracing breath of fresh air. Wearing only the boots, you slowly trot out of the hot sauna, and then out of the Station into the darkness and the invigorating -100°F air. You loop around the South Pole, and then head back inside again. The whole process takes a few minutes.

In the act of slowly circling the South Pole, you are in fact circumnavigating the globe of the world rather quickly. You're also crossing all 24 Time Zones – but the jet lag is minimal!

NAKED IN ANTARCTICA

The prospective 300 Club members are discouraged from wearing underwear while in the sauna, because it will get soaked with sweat (and the successful entrants claim that it's more fun to be naked). You definitely do not want your sweat to freeze instantly as it hits the -70°C air.

One danger is that a thin sheet of ice on your boxer shorts could give you a nasty cut.

A nastier and more subtle problem is that the ice has both a higher thermal capacity (can store more "coolth") and a higher thermal coupling (can suck heat from your skin) than air. So ice at -70°C touching your skin will cool you much faster than air at -70°C. This could cause hypothermia.

The 300 Club experience is more exhilarating than enjoyable.

You do it to say you've done it. You also do it as part of the social bonding needed to get through the physical and mental difficulties of surviving the eight-month-long winter – with four months of darkness. It gives the 300 Club members a shared identity in having done something both exotic and dangerous. But it does not make you feel physically good. Back in 2006, the then-South Pole Winter Site Manager, Andy Martinez, said, "It just felt like somebody was hitting me with a tennis racket full of needles." On the day he did it, there was a slight wind. So the Wind Chill Factor made

VERY COLD = HARDLY ANY WIND

By the way, for the temperature at the South Pole to get down to -100°F, there has to be hardly any cloud. This is because of the so-called "Radiative Effect" – to get really cold, the surface has to be able to radiate its heat to a cloudless sky. (Clouds reflect the heat from the ground back down again.) It turns out that a cloudless

the -75.5°C air temperature "feel" like -97.7°C. (Mind you, there's a lot of debate about the accuracy of "Wind Chill Factors" down at extremely low temperatures.)

SOUTH POLE HIGH

Living at the South Pole is living at altitude.

It's 2,835 metres above sea level. About 2,700 metres of that is ice – the remainder is rock.

But the "effective" altitude is somewhat higher.

First, the air pressure at the Pole is lower than at the equator. Second, it's always cold. The extremes recorded so far are -82.8°C (-117°F) in the depths of winter, and -12.3°C (9.9F) in the "heat" of summer. This relentless cold also makes breathing harder.

SURVIVING 300 DEGREE CHANGE

So how do you survive the extreme heat of the 200°F (about 93.3°C) sauna? After all, you can cook an egg or a steak at that temperature.

The critical difference is that we humans can sweat.

Provided that you are hydrated enough, your skin will pump out sky at the South Pole is usually associated with little or no wind.

This means that when the thermometer hits -100°F, there is usually very little wind. This is good. The 300 Club event is just survivable in "suitable" conditions. Any significant wind would remove too much heat from the naked flesh – leading to the event being "incompatible with life".

There have been injuries. On one occasion, a male and female simultaneously suffered minor frostbite of the foreskin and nipple, respectively.

enormous quantities of salty water. As this sweat evaporates, your skin cools down. Mind you, once you get dehydrated and can no longer sweat, you will rapidly overheat.

But there's another factor. The sauna has to be a dry sauna, so that the sweat on your skin can evaporate. If the air in the sauna is heavy with water, then your sweat can't evaporate – and again, you'll overheat.

What about the extreme cold of less than –100°F (about -73.3°C)?

When your skin gets exposed to this low a temperature, your body desperately tries to trap or conserve all of its heat. So the blood vessels in the skin close up very suddenly, and a lot of warm blood is immediately shunted or relocated internally. Unfortunately, your heart is exposed to a massive overload of blood – but only for a few seconds. Even so, you need a healthy heart with a lot of reserve capacity to do this trick.

OCC. HEALTH N' SAFETY

Breathing through a scarf tends to trap warm moist air in your mouth and throat – and also tends to reduce how far the super-cold air makes it down your upper airways. But most people don't use a scarf.

A mate of mine, Darryn Schneider, was down at the South Pole, installing and working on the underground AMANDA Telescope (Antarctic Muon and Neutrino Detector Array.) It was the precursor

MARRIED ON DIFFERENT DAYS

In February 1986, Sasha Zemenak (winter cook) was married to Lee Schoen (station manager) at the South Pole. They were married to each other on different days.

How come?

During the ceremony, Lee stood on the February 1 side of the International Date Line, while Sasha stood on the February 2 side.

to Ice Cube Neutrino Telescope, which Darryn also worked on. Ice Cube Telescope is some 94 strings of optical sensors, buried vertically in the ice. They run from about 1.45 to 2.45 kilometres below the surface of the ice, and they detect neutrinos (check out Wikipedia).

Back in early July, in the year 2000, the temperature dropped to -100°F. So Darryn Schneider tried out for the 300 Club. He survived the +200°F sauna, toweled off the sweat, and walked the length of the tunnel to the outside world – which was waiting to whack him at -100°F. Then he made his big mistake. Instead of walking the hundred-or-so metres to the South Pole, he ran. After a very short while, he began to feel a little chilly, and sensibly came back inside into the warmth. And then it hit.

He began coughing. Darryn was lucky, because he had only mild symptoms - a cough for the rest of the day. Some of the other people who ran for the 300 Club event had a cough for a few days. They had what is called "frostbite of the lungs".

So slow and steady, and with a decidedly thick skin, maybe will get you through the 300 Club, if you're game ...

(That could lead to "complications" in the paperwork.)

The wedding rings were made from small Caterpillar bulldozer thrust bearings.

SOUTH POLE HABITATION

The Amundsen-Scott South Pole Scientific Station honours both Roald Amundsen (reached South Pole, December 1911, all of his team survived) and Robert F. Scott (reached South Pole, January 1912, all of his team perished).

The first permanent human habitation at the South Pole was set up by the US, as part of their Cold War strategy, during 1956-1957. It was abandoned in 1975, and is now called Old Pole.

It was replaced by the Dome, which was built over the summer of 1974-1975. It was a geodesic dome some 50 metres wide and 16 metres high, with various buildings inside. It was abandoned in 2003.

The third habitation has adjustable legs, to stop it from being buried in snow. The summer population (November to February) is about 200, dropping to about 50 over the winter.

13

THE KILLER DRILLER

FOR THOUSANDS OF YEARS, THERE HAVE BEEN TANTALISING REPORTS THAT DRINKING THE WATERS OF CERTAIN HOLY RIVERS (SUCH AS THE GANGES AND YAMUNA RIVERS IN INDIA) COULD CURE INFECTIOUS DISEASES SUCH AS LEPROSY, CHOLERA, DYSTENTERY AND THE BUBONIC PLAGUE. (THIS WAS AS LONG AS YOU DIDN'T GET CHOLERA, ETC., FROM DRINKING THE WATER IN THE FIRST PLACE.)

It was tantalising because if this was a real effect, maybe it would mean that there was another way – besides antibiotics – to kill bad bacteria. It turns out that this effect is real. In fact, some of us have been using it for nearly a century. This "new" option is a virus that kills half the bacteria on Earth every two days. Say howdy to your new best friend forever, the "Bacteriophage" – usually shortened to "phage". The name literally means "bacteria eater".

ANTIBIOTIC-RESISTANT BACTERIA

Phages can help us fight bacteria that have evolved resistance to antibiotics.

We have used antibiotics since the 1930s to treat bacterial infections.

Like Global Warming, Evolution is real. Bacteria have evolved resistance to the antibiotics we use to kill them. We then invented new antibiotics. But in response the bacteria evolved resistance to the new antibiotics – and the cycle repeats ad infinitum. (I discuss this in the story "Bacteria Burger", in my 31st book, *Brain Food*.)

The problem is that we are already running out of new antibiotics – but bacteria won't run out of Evolution. Bacteria have evolved many different pathways to become resistant to our antibiotics.

Some bacteria evolved to secrete enzymes that break down the incoming antibiotics before they got too close. If the antibiotics worked by breaking down the bacterial cell wall, bacteria evolved new, impervious cell walls. If the antibiotics could penetrate the bacteria's cell wall and do their damage inside, bacteria evolved pumps to eject the antibiotics.

Each year in the USA, about two million people are infected with bacteria that are resistant to antibiotics. The direct result is that 23,000 people die – and the costs to the economy are about US$20 billion annually. Some 30 per cent of severe *Streptococcus* pneumonia infections are resistant to multiple antibiotics.

It costs at least US$1 billion to develop a new antibiotic. But every time we do, the bacteria it targets will evolve to resist it. Antibiotics can't evolve – they are just a chemical.

If only we could find a creature that is both able to kill bacteria when it encounters them – and also capable of evolving to follow, find and destroy them as the bacteria try to evolve their way out of danger.

Luckily, there is such a creature – the virus known as the bacteriophage. Bacteria and phages have been in a vicious arms race for millions of years.

BACTERIOPHAGE 101

From a Molecular Biology point of view, phages are "self-replicating nanoparticles".

Phages deal with bacteria in two main ways. They either use and destroy them, or they use and don't destroy them.

They infect a bacterium, and take over its DNA/RNA so that it makes a hundred or so more phages. They are usually very specific about what bacteria they attack – so they leave human cells unharmed.

Your typical phage virus looks a bit like a Moon Lander – a tall skinny body on top of a bunch of spindly legs. Phages can't move independently – they rely on random encounters due to currents in blood, lymphatic circulation, rivers, sewage, irrigation waters, etc.

The phage lands on a bacterium, legs down. The legs then bend, lowering the tiny body of the phage to make contact with the much larger body of the bacterium. But phages are very specific about which receptors (chemicals) they will attach to on the bacterium. This is the basis of the ability of the phage to attach to one particular bacterium, but not another.

The tail of the phage then penetrates into the bacterium and injects some of its genetic material. This bacteriophage DNA then forces the host bacterium to start making more bacteriophages – sometimes as quickly as within 15 minutes.

The bacteriophage can exit the bacterium in many different ways, depending on various factors. Sometimes the bacteriophage forces the host bacterium to continuously dribble out more brand-new bacteriophages. Sometimes the new bacteriophages get released in batches, in little buds or capsules.

But sometimes phages actually destroy their host bacterium. They then burst out into the local environment in a giant onslaught. They erupt by demolishing the tough cell wall of the bacteria. The internal pressure of the bacteria literally bursts open the cell wall. This explosion releases a hundred or so bacteriophages to attack more bacteria.

The Demolition Machines used in this case by the bacteriophage are called "endolysins". Scientists are now trying to understand how endolysins work. (When they do, it will give us another weapon against antibiotic-resistant bacteria).

The net result is that bacteriophages kill about half the bacteria on Earth every two days. On the other hand, bacteria replenish their numbers very quickly – so we still have pretty well the same number of bacteria all the time.

PHAGE THERAPY

There are many ways to take phages into our bodies.

The currently most popular way is to lay them as a bandage onto a chronic wound infection, such as an ulcer that won't go away. One such product from the Eliava Research Institute in Georgia is a biodegradable wound dressing that is impregnated with a local anaesthetic, an antibiotic – and six specially chosen phages.

But phages can also be taken orally, rectally, and by nasal spray, injection, mouthwash, eye drop and tampon.

They are removed from our bodies by two major pathways – cells of the reticuloendothelial system (check it out on Wikipedia), and antibodies.

But sometimes they can find a niche, and persist inside us indefinitely.

THE BAD SIDE

Everything has its down side, and that holds true for phages also.

One current problem with phages is the time delay. They are each highly specific to just a few bacteria. This means that you have to identify the attacking bacteria very accurately. At the moment, this can take 48 to 72 hours. But new technologies are becoming available that will reduce this time to a few hours, or even minutes.

Another quite sinister potential problem relates to the phages that do not kill bacteria, but instead live with them in a mutually beneficial symbiotic relationship. The first issue, of course, is that they don't kill the nasty bacteria. The second issue is more subtle. What if they set up a relationship with a nasty bacterium that is already resistant to some antibiotics? It's possible that the phage can pick up this antibiotic resistance, then infect other bacteria that are not yet resistant to antibiotics – and make them resistant.

And, rarely, a phage can carry genes that accidentally increase the virulence of a bacterium, rather than kill it. For example, the phage might stimulate the bacterium into making nasty toxins. This is how the cholera toxin evolved.

Then there's a simple mechanical problem – sometimes phages don't work. Some bacteria, the so-called Gram-negative bacteria, have an outer cell membrane. This can act as an impenetrable barrier to the phages.

GOOD EXAMPLES

So now that I've got the obligatory Let's Not Get Irrationally Exuberant Over Phages section out of the way, we can go ahead and look at their potential for good.

You've probably heard of Golden Staph – *Staphylococcus aureus*. In many cases, it's resistant to the most powerful antibiotics we have, such as methicillin. In one study, methicillin-resistant *Staphylococcus aureus* was exposed to a tiny dose of 10 million phages. Within 10 minutes, over 99.9 per cent of the bacteria had died.

PHAGES ARE ALL DIFFERENT

We currently know of 19 "families" of phages that can infect bacteria and archaea. (Archaea are similar to bacteria in shape and size, but much older, run on a different biochemistry, and can withstand extreme environmental conditions. They are sometimes called "extremeophiles".)

Here's another example.

The *Pseudomonas aeruginosa* bacterium can cause hard-to-shift ear infections. It's especially hard to treat with antibiotics, because some of the species of *Pseudomonas aeruginosa* have evolved to wrap themselves inside a "biofilm". (The biofilm concept is discussed in my 17th book, *Flying Lasers, Robofish and Cities of Slime*, in the story "Cities of Slime".) This biofilm is made of layers of proteins and sugars, and blocks the ingress of antibiotics by a factor of about 1000. You guessed it – a single dose of a phage therapy called Biophage-A has cleared up long-term antibiotic-resistant ear infections.

PHAGE ADVANTAGES

It was originally seen as a disadvantage that each type of bacteriophage evolved to attack only a very limited range of bacteria. But this turns out to have advantages.

If you take an antibiotic to treat an ear infection, it will usually kill the bad-guy bacteria responsible for your pain and misery. Unfortunately, it will also accidentally kill many of the essential and friendly bacteria that live in your gut, and elsewhere in and on your body. (Collateral Damage is the current term.) Unfortunately, once the balance of the bacteria in the gut is changed, in some cases, very nasty bacteria such

> Phages' genetic material can be RNA (two families) or DNA (17 families). Their DNA can be linear (eight families) or circular (nine families). Nine of the families infect only bacteria, another nine infect only archaea, while just one infects both.

as *Clostridium difficile* begin to flourish. In 2014, *Clostridium difficile* killed 14,000 people in the USA. (This is also discussed in my 31st book, *Brain Food*, in the story "The Woman Whose Life was Saved by a Poo Transplant").

The beauty of bacteriophage therapy is that it will leave these friendly bacteria alone. Bacteriophages, as personalised medicine, are more like a guided missile than a nuclear bomb.

Phages have a major difference from antibiotics. Immediately after you take antibiotics, their level in your body drops. But immediately after you take phage therapy, their level increases. However, once the phages have done their job, because there are no bacteria left for them to breed in, they tend to vanish.

Another advantage relates to what bacteria do very quickly – evolve. Suppose the bacteria you're treating have evolved to be resistant to one type of bacteriophage. Luckily, there is always another slightly different bacteriophage just around the corner. Nature has a virtually inexhaustible supply of them. So you just use another bacteriophage to treat the patient. Indeed, we can now use modern Molecular Biology to quickly match up a bacteriophage with the intended target.

There's another unexpected advantage. With antibiotics, you have to deliver them to every single person you wish to treat. But phages, like bacteria, can jump from person to person. So if you're not in a hurry, you can simply treat a few people, and let the phages do the walking for you.

In general, once their main target (for example, a nasty bacterium) has gone, the overwhelming majority of phages fade away. But in many cases, a tiny minority hang around. In this situation, just like vaccines, phages can give long-term immunity.

SHEEP PHAGES KILL *E.COLI* O157:H7

Back in 2003, Dr Andrew Brabban from Evergreen State College in Washington State was testing antibiotics. He had an unexpected problem – which gave him a wonderful solution.

He had hoped to use these antibiotics to vanquish a particularly nasty bacterium called *E. coli* O157:H7. The bacterium *E. coli* is normally harmless to us, but the specific version *E. coli* O157:H7 is a major cause of food poisoning, sometimes causing kidney failure. It has been implicated in many human deaths. Annually, it hospitalises some 2100 people in the USA alone.

Many livestock, including sheep, can carry it without suffering any symptoms. But when the animal is slaughtered, the *E. coli* O157:H7 can pass into the meat. It can also infect manure.

Dr Brabban deliberately infected sheep with *E. coli* O157:H7, so that he could test his antibiotics. "Unfortunately", as quickly as he infected the sheep, the bacterium vanished.

You guessed it – the sheep were carrying a phage that turns out to kill 16 of the known 18 toxic strains of *E. coli* O157:H7. This discovery may help us deal with *E. coli* O157:H7.

VIRUSES RULE

Viruses are everywhere – from two kilometres under the Sahara Desert, to polar lakes and acidic hot springs. The number of virus particles in each millilitre ranges from 60,000 on the ocean floor of the Barents Sea to 254 million in the surface waters of Lake Plussee in Germany.

It is hard to express how many virus particles there are on Earth. One way is in straight Scientific Notation – 10^{31} particles (that's 10 million million million million million). Another way is that if they were laid end-to-end, the virus particles would reach, not just to the Moon, or to Pluto, or to the Andromeda Galaxy, some two million light years away. No, they would reach 125 times further – 250 million light years.

Viruses are Nature's top Evolutionists. They continually and randomly combine with any DNA they run across, whenever they infect any new cell. As a result, some 10^{24} new viruses – a million million million million – are being created each second. The overwhelming majority won't survive (as they are not compatible with life), but some will.

PHAGES RULE

It appears that there are more undiscovered genes on our planet (mostly belonging to phages) than in all the other life forms combined. They are the most numerous life forms on Earth. (But is a virus "alive"?) There are about 100 million phages in each gram of dirt, or each millilitre of water.

Phages insert some of their DNA into their host bacterium – either temporarily or permanently. As a result, they have helped some bacteria make death-dealing toxins. (Nothing personal, Humanity – it's just Evolution.) That's where the primary toxins of botulism, cholera, diphtheria and scarlet fever came from.

Phages helped turn harmless *E. coli* into the potentially fatal *E. coli* O157:H7. The difference is a million base pairs of DNA that seem to have come from 24 separate phage invasions. As another example of how phages can make a bacterium truly nasty, consider the bacterium *Pseudomonas aeruginosa*. It uses two sets of phage genes (which it got from some unknown phage) to attack and kill its competitors.

Phages also play a role in Climate Change/Global Warming. By killing and controlling the numbers of enormous populations of bacteria, phages affect the Carbon Cycle. They also prey on marine algal blooms. This means that they influence how much dimethyl sulfide the algae release – which means they influence cloud formation. (Read more in the story "Ozone at Seaside", from my 26th book, *Please Explain*.)

BACTERIA USE PHAGES AS WEAPONS

It seems that some bacteria have seen the light. They have taken over phages, to use them as weapons against other bacteria, or much larger creatures such as insects. They release these phage-drones from their little single-celled bacterial bodies.

Pseudomonas aeruginosa attacks other bacteria with modified phages. These phages do more than make a hole in the target bacteria – they deliver a fatal toxin.

The bacterium *Serratia entomophila* attacks New Zealand Grass Grubs (small beetle larvae), again using modified phages. Here, the bacterium's aim is so specific that the phages do not affect other insects that are closely related.

But there's a downside to this wonderful phage weapon – it's a suicide mission for the bacteria. As the phages are released, they are so large that they fatally burst open the bacteria. Why would bacteria engage on a Kamikaze Mission? It turns out that like us, many bacteria are highly social. They sacrifice themselves for the common good.

Hopefully we can learn how to make modified phages, so we can better target specific bacteria.

PHAGE POWER

Phages have been used to make, and store, electrical power. Let me introduce you to one specific phage, known as M13. It has a rather unusual long and skinny rod-like shape – seven nanometres across, but 900 nanometres long.

Thanks to its shape (and the Piezoelectric Effect – you can look it up on Wikipedia), the M13 phage can convert Mechanical Energy to Electrical Energy – and vice versa. It has powered a 10-square-centimetre LCD screen. There are plans to make it power an internal nanoscale medical device by collecting Mechanical Energy from a beating heart, or a pulsating artery.

The M13 phage has also been used to store electrical power. In this case, Professor Angela Belcher, a biological engineer at MIT, used a marine snail as her inspiration. Marine snails use proteins to guide their incremental laying-down of calcium carbonate to make a shell. A phage like M13 is just genetic material inside a protein coat. Protein helps the phages do a fantastic job of assembling themselves inside a recently invaded bacterium.

Professor Belcher worked out how to get M13 to assemble something useful to humans – electrodes for a Bio-Battery.

She modified M13 so that its protein coat would bind with materials she wanted to use as electrodes in a lithium-ion Bio-Battery – materials such as cobalt oxides and iron phosphates. When M13 "built" these nano-wire cathodes, they turned out to have a spiky surface. This was great – the extra surface area meant they could carry more current.

The phage-based Bio-Battery works – but will currently deliver only hundreds of cycles, not thousands, so the battery isn't usable for long. However, it's early days.

PHAGES PROTECT GUT

In your body, only 10 per cent of the cells carry the DNA you got from your parents. The remaining 90 per cent are invaders – bacteria and the like. They make up our Gut Microbiome. (See "The Strangers Within", in my 31st book, *Brain Food.*) We need them, and we want them close – but not too close. If they get too close, these foreign bacteria (and the like) can invade our cells that make up the gut lining.

There is a thick coating of slime (mucus, or mucin glycoproteins if you want to get technical) on the inside wall of the gut. It helps keep the Gut Microbiome at bay. But what is it that is keeping them close, but not too close?

Again, you guessed it – phages. If the phages are not there, the bacteria move in, invade our gut cells, and cause inflammation.

VIRUS – ALIVE OR NOT?

Is a virus "alive"?

Depends on your definition of "alive".

Suppose you get a swimming pool full of sterile, pure water. You add the basic foods needed for life – fats, proteins and carbohydrates. Then add some "fairy dust" – the minerals needed for life, such as phosphorus, iron etc.

Add one single bacterium. It will grow and multiply, and eat all the food and generate trillions of bacteria. So the bacterium is alive.

Let's start again with the same pool with basic foods and minerals – but completely sterile. There is not one single living creature inside it – no bacteria, no algae, no fish, nothing.

Add one single virus particle.

Nothing happens. You don't get trillions of viruses. Is the virus "alive"? Not by this test.

But then add some bacteria. The viruses invade the bacteria, take over their DNA/RNA, and make more viruses.

So a virus can't breed with just the basic foods and minerals – as bacteria and we humans do. It needs one more factor in addition – a living creature that it can infiltrate.

Do I think that a virus is "alive"? On Mondays, Tuesdays and Wednesdays I think "yes", on Thursdays, Fridays and Saturdays I think "no" – and on Sundays, I don't know.

14

SURPRISING SUN

SOMETIMES THE MOST FAMILIAR THINGS CAN BE THE MOST SURPRISING. AND NONE IS MORE SURPRISING THAN THAT BIG BALL OF NUCLEAR FIRE IN THE SKY ... THE BIG FELLA THAT KEEPS US ALL ALIVE, THE SUN.

HOW DOES THE SUN BURN WITHOUT OXYGEN?

The Sun's burning is not a chemical reaction – it's a nuclear reaction.

Chemical burning always involves the electrons getting shifted around (say, from one atom to another). And an oxidiser – usually oxygen – is involved. The nucleus doesn't get involved. It just goes along for the ride.

Nuclear reactions directly and

So how about this for a surprise? Our Sun has long-lost siblings – and it rains on the Sun.

SUN 101

First, let's start off with the basics. The Sun is about 150 million kilometres away, and about 1.4 million kilometres in diameter. So, compared to Mother Earth, our Sun is about 109 times bigger, and has a mass about 330,000 times greater. So far, in its 4.6-billion-year life, the Sun has burnt about 100 times the mass of the Earth, and turned it into energy. But the Sun is so big that this loss is just a tiny 0.03 per cent of its total mass.

Thanks to nuclear burning, when the Sun is directly overhead, it dumps about one kilowatt of power on each square metre of the Earth. (By way of comparison, it takes about a kilowatt to run a microwave oven.) That power is about 50 per cent infrared heat, about 40 per cent visible light and about 10 per cent ultraviolet light.

At this stage in its evolution, the Sun is made up of 74.9 per cent hydrogen and 23.8 per cent helium, with much smaller trace amounts of the elements oxygen, carbon, neon (about 0.2 per cent) and iron (about 0.2 per cent). As the Sun ages, these percentages will change.

The Sun is not made from any of the three States of Matter that we

fundamentally involve the nucleus of the atom. (That's how we get the names Nuclear Power, Nuclear Energy and Nuclear Weapons.)

Every second, the Sun burns around 620 million tonnes of hydrogen, and turns it into about 616 million tonnes of helium. Every second, there is a shortfall of about four million tonnes – and most of this is converted into energy, as described by Einstein's famous $E = mc^2$ equation.

are mostly familiar with – solid, liquid or gas. No, the Sun is made of the fourth State of Matter: plasma.

You might have seen a model of an atom that looks similar to a mini-solar system. It has the heavy nucleus in the centre, with a cloud of much lighter electrons surrounding it.

A plasma is just a bunch of atoms that are so hot that the electrons have broken free and are no longer tied to the nucleus. The electrons float freely through a sea of nuclei. This is "plasma" – what the Sun is made from. The nuclei of the atoms of hydrogen, helium and other atoms are all present – but their electrons wander freely.

SUN'S LOST SIBLINGS

So now you know the background, let's consider the Sun's long-lost siblings. Right now, our Sun is pretty lonely – over four light years from its nearest neighbour, Alpha Centauri.

But its birth wasn't solitary.

Like most stars, our Sun condensed from a vast molecular cloud of dust and gas (mostly hydrogen and helium). 4.6 billion years ago, the Sun was most probably born in a cluster of perhaps 1000 other stars – all from the same cloud.

The astronomers went looking, and found some stars with the same trace elements, in the same ratios, as our Sun.

They also did a kind of "Galactic Archaeology", using complex mathematics to "rewind" the motions of some candidate stars over the last 4.6 billion years – back to that primeval cloud.

There are many reasons why the stars from that original cloud could have separated. Over that long time, our galaxy, the Milky Way, has rotated on its own axis about 18 times. We're about 25,000 light years from the giant Black Holes in the centre of the Milky Way – about two thirds of the way out to the edge. So that's a lot of "roads less travelled" for the wandering stars to go their separate ways.

Another possibility is that the gravity of the gas cloud that created the stars could hurl them randomly into the galaxy at large. Interactions between newly created stars (inside the gas cloud) could result in the ejection of other stars. Alternatively, passing stars could pull out a star from inside a gas cloud.

Imagine the Galactic Archaeology problem to be a bit like dribbling a drop of food dye into the Pacific Ocean, and then coming back a year later and trying to find all the atoms that made up that original drop.

You can see that the mathematics was not trivial.

Against all odds, the astronomers have found what is almost certainly one of our Sun's siblings. It's about 110 light years away in the constellation Hercules, and has the rather unromantic name of HD 162826. It's a little bigger than our Sun, and a little hotter – but it's the same age, because it was born at the same time. (We currently don't know if, like our Sun, it has planets.)

RAIN ON THE SUN

So the Sun has at least one sibling that we know of, but how can it possibly have "rain"? After all, down in the Sun's core where hydrogen is turned into helium, the temperature is around 15,000,000°C. Even on the surface, it's about 5500°C.

OK, the "rain" is not made from water, but it does fall down just like rain does, here on Earth.

On Earth, there is the so-called Water Cycle. Water vapour forms (say, over the oceans), and it rises upwards to make clouds. As these clouds rise further to a few kilometres of altitude, they cool – and they can condense into little water droplets, which then fall down as rain. These droplets are a few millimetres across. (Note that in this cycle, water undergoes changes of state, with liquid first turning into gas, then turning back into a liquid.)

On the Sun something similar happens – but it's with plasma, and the scale is so much bigger. The Sun doesn't really have a hard surface, but there is a boundary between its main body and its atmosphere (which is called the "corona"). This Coronal Rain was predicted some four decades ago, but only recently have our more advanced telescopes seen it.

Via various processes, the super-hot plasma from under the Sun's surface gets squirted to a height of about 60,000 kilometres, cools to about 60,000°C, and then falls down again. One difference from Earth's rain is that there is no change of state – the plasma remains as plasma. The other big difference is the size of the droplets – they're about 100,000 square kilometres, or roughly the size of Tasmania.

And when they fall, they plummet down to the Sun's surface at up to 50 kilometres per second. This is surprisingly slow – it "should" be a lot faster, thanks to the Sun's enormous gravity. But it appears that the falling "raindrops" are cushioned by a huge layer of hot gas – which slows them down.

MORE MYSTERIES

Despite the surprising things that we do know about our closest stellar neighbour, there's still so much we don't know. For example, we don't know why the Sun's atmosphere is over a million degrees hotter than the Sun's surface.

And we don't know where the energy that makes it so hot comes from. (One theory is that it's related to the Coronal Rain.) But in 2017, the European Space Agency will launch the Solar Orbiter spacecraft. Not only will it get within 45 million kilometres of the Sun, but it will also give us our first long-term views of the Sun's North and South Poles. And the following year NASA will launch its Solar Probe Plus spacecraft, which should get within a cool six million kilometres of the surface of the Sun.

That should be close enough to throw some light on new surprises . . .

SUN WILL SWALLOW INNER PLANETS

The Sun is about 4.6 billion years old, and is roughly halfway through its life in its current "size". In another seven billion years or so, the Sun will evolve into a Red Giant, and expand – enormously (maybe to 100 times bigger).

It will get so huge that it will swallow the inner rocky planets of Mercury and Venus, and possibly even Earth.

Today . . .

The Sun

Mercury
0.38 AU

Venus
0.72 AU

Earth
1 AU

Mars
1.52 AU

7.588 billion years from now . . .

The Sun as a red giant
0.9 solar mass

Earth
1.1 AU

Mars
1.69 AU

7.59 billion years from now . . .

The Sun as a red giant
0.8 solar mass

Mars
1.9 AU

Surprising Sun   < 113

SATURATED FATS

IF YOU THINK THAT CLIMATE CHANGE IS "CONTROVERSIAL", TRY SATURATED FATS. IN CONTRAST TO CLIMATE CHANGE, AT THE TIME OF WRITING, SOME OF THE SCIENCE ABOUT THE HARMFUL EFFECTS OF SATURATED FATS IS STILL SOMEWHAT DEBATABLE.

It seems that "chain length" is part of the story.

In general, Long Chain Saturated Fatty Acids (coming mostly from animal foods) seem to be definitely bad for your health. They are associated with increased LDL cholesterol, increased risk of cardiovascular disease, and poor metabolic health. These associations are strong both in real studies in the world where we live, and in laboratory "in vitro" studies.

However, the jury is still out with regard to the Short and Medium Chain Fatty Acids, such as those in dairy and coconut.

Many studies show that the effects of the Short and Medium Chain Fatty Acids are "neither good, nor bad", or might even be protective. (This might be related to their different mechanism of absorption and metabolism by the body.) Indeed, dairy overall is associated with health benefits – dairy foods contain calcium and protein, have low Glycemic Index (GI) values, etc. But the full answer is not in yet.

There is still much research to be done to understand the effects of different chain lengths of fatty acids – whether they are saturated or unsaturated.

Let me remind you of the Dunning–Kruger Effect (covered in the story "Clueless" in my 34th book, *Game of Knowns*). This Effect explains how people who are incompetent in a particular field don't know how incompetent they are. On the other hand, experts (who are deeply competent) know the limits of their knowledge.

Let's apply the Dunning–Kruger Effect to the field of Dietetics and Nutrition.

Experts – professors and fellow academics, who have spent decades doing actual research – openly acknowledge the ambiguities in what we know about Saturated Fats. They understand the subtleties and uncertainties of research at the cutting edge.

However, what about those "experts" who got their "expertise" by "passing" a 20-minute online questionnaire on "nutrition", and now have a colourful "certificate" (that they printed at home)? Surprisingly, they can be supremely confident that saturated fats are always good for you, and can't ever be bad. They share similarities with people who believe in the No Carb diet fads, and who claim that all fats are good and that all carbs are bad. They don't even consider the type of fat or carbohydrate.

So a little knowledge can be harmful, if you don't know what you don't know...

Saturated Fats < 115

16

COCONUT OIL AND WATER

THE LAND OF NUTRITION IS NOTORIOUS FOR THE INCREDIBLY LARGE NUMBER OF FOOD FADS THAT COME AND GO. (SURE, YOU'LL FIND CRAZY IDEAS IN PHYSICS AND CHEMISTRY TOO – BUT FOR SHEER VOLUME OF NEW TRENDS, VISIT FOOD LAND.)

A currently popular fad is based on that "Fruit" of the Tropics, the coconut. (Wheatgrass, goji berries and chia seeds are soooo 2010.) Proponents claim that coconut oil and coconut water will do everything from beautifying your skin to boosting your immunity and metabolism and aiding your digestion. Promoters also claim that coconut products have only good results, and never any bad side effects.

But everything has side effects, so I'm already dubious about these claims.

COCONUT HISTORY

The coconut was very useful in the early days of human exploration. I mean *really* early – some 70,000 years ago, when we walked out of Africa in the depths of an Ice Age and then spread through the tropics.

In one neat little package, the coconut is a combined source of both nutrition (fat, protein and carbohydrate) and water. It made it much easier for us to travel and colonise the Pacific Rim, coastal India, Africa and South America.

Today, the coconut is cultivated on some 100,000 square kilometres across some 86 tropical countries. The annual crop is about 62 million tonnes.

COCONUT WATER 101

Coconut water is the clear liquid inside a coconut. It's claimed to help you lose weight, and improve your skin tone and digestion.

It is moderately rich in potassium and also contains small amounts of minerals such as magnesium, calcium and phosphorus. However, both the banana and the potato carry roughly the same amounts of potassium as coconuts – but you don't see potatoes being endorsed by celebrities and sold as the next sure-thing Superfood. So these minerals surely can't be the basis of their health claims.

Coconut water has been endorsed by celebrities from Lara Bingle to

Madonna. When coconut water was first introduced in the USA as a Superfood in the early 2000s, it was claimed to have the power to fight kidney disease, osteoporosis and viruses. However, dieticians shot down these incorrect claims. Other false claims for "Mother Nature's Sports Drink" include the ability to cure hangovers, slow the ageing process, and simultaneously lower your weight, blood pressure and cholesterol.

TYPES OF COCONUT OIL

It turns out that there are two major types of coconut oil, depending on how the coconut is processed.

The better grade is the so-called virgin coconut oil, which is processed from fresh coconut flesh or coconut milk. It supposedly contains more of the good stuff. Many of the claims made about "regular" or refined coconut oil are actually related to the much more expensive, and much less common, virgin coconut oil.

COCONUT WATER – REHYDRATION FLUID?

The worldwide financial spend on coconut water in 2013 was nearly half-a-billion dollars – sold on supermarket shelves, and in yoga studios and gyms. It is being heavily pushed as a rehydration liquid for athletes and lovers of natural food.

Sure, when you sweat, you lose water, sodium and lots of minerals. But studies have shown that coconut water is about as good at rehydrating you as generic sports drinks or – wait for it – water out of the tap. The false marketing claim that it is superior to sports drinks as a rehydration liquid was withdrawn in the USA after a 2011 class action lawsuit.

But what about the elite athletes who push themselves for several hours every day – and generate huge amounts of sweat?

SUPERFOOD SPUDS?

An average-sized potato has roughly the same amount of potassium (900 milligrams) as a bottle of coconut water (800 milligrams).

But a potato is also high in fibre, and also carries your recommended daily requirement of Vitamin C. And it has no sugar or fat.

Perhaps we should save the prefix "Super" for the words "Man", "Annuation", and "Conductor" . . .

Sweat contains about 10 times as much sodium as potassium. But coconut water runs the other way – it has about one half to one third as much sodium as potassium. In coconut water, the ratio of sodium to potassium is in the wrong direction – and wrong by a factor of 25. Coconut water simply does not have enough sodium to do a good job with sodium replacement for a high-performance athlete.

But wait, there's more. If you drink large amounts of coconut water to get sufficient sodium, you'll soon realise that coconut water has a laxative effect. To put it mildly, having diarrhoea is sub-optimal if you are trying to replenish your bodily fluids. You could also have taken onboard too much potassium (potentially leading to heart problems).

Another problem for elite athletes: because coconut water is not formulated in a factory, its ingredients can vary enormously from batch to batch. (Variety is good for a regular non-athlete. But elite athletes are shoving huge quantities of foods and liquids though their body, so they need to know exactly what they are taking in.)

Coconut water has less sugar than soft drinks (about 2–3 grams sugar per 100 millilitres versus 7–13 grams sugar per 100 millilitres). This is good. On the other hand, it has much more sugar than tap water, and much less fluoride. This is bad. Tap water should be your first choice as a thirst quencher. But as a refreshing occasional drink, coconut water is fine – especially if you get it straight from a coconut, not a bottle.

MINERALLY

The definition of a "mineral" is vague.

It can mean something that has been mined from the ground.

It can also mean an element or a chemical with a specific chemical formula – usually inorganic, and not derived from a living creature. So the word "mineral" can be applied to food ingredients, such as calcium and phosphorus.

Just don't waste your money filling your pantry with it in the false belief that it is health-giving.

COCONUT OIL

What about the health claims for coconut oil?

Like coconut water, it also has celebrity endorsers ranging from Olympic Champions to movie stars such as Angelina Jolie and Miranda Kerr (who is quoted as eating four spoonfuls every day). It's supposed to keep hair shiny, and skin softly textured and beautifully moisturised. Some of its other loudly trumpeted health benefits include controlling sugar cravings and weight, as well as relieving stress and boosting your immunity. Not unexpectedly, there is no compelling evidence for these claims.

Surprisingly, coconut oil is very rich in saturated fats. (See "Saturated Fats" on page 114.) This is quite different from practically all the other oils that come from plants – overwhelmingly, they are not saturated. Coconut oil is about 91 per cent saturated fats and only six per cent mono-unsaturated fats. This is virtually the opposite of olive oil – 14 per cent saturated fats and 72 per cent mono-unsaturated fats. The benefits of olive oil (for example, as part of the Mediterranean Diet) are well recognised – from both medical and historical perspectives.

COCONUT OIL SMOKE POINT

One oft-quoted advantage of coconut oil over olive oil is that it has a higher Smoke Point – the temperature at which the oil begins to emit a fine blue smoke.

There are two problems here with this claimed advantage.

First, you should never heat any oil to its Smoke Point. Bad chemical

However, from a storage point of view, saturated fats have advantages. They make coconut oil resistant to oxidation, which can turn it rancid – so you can store it for a few years before it goes off. This is good for the manufacturer and retailer.

But from a health point of view, saturated fats in large amounts can be dangerous. They are very strongly associated with bad blood cholesterol levels and heart disease. This is the overwhelming majority view of bodies such as the United States Food and Drug Administration, the World Health Organisation, the American Heart Association, the American Dietetic Association and many other professional medical and dietetic organisations.

A typical comment about saturated fats comes from Professor Mike Gordon, a food chemist at Reading University. He says, "Coconut is 50 per cent made up of a fatty acid called lauric acid which if consumed to excess can contribute to increased risk of coronary heart disease. I would be very cautious about advising people to consume large amounts of coconut oil."

Of course, there is an opposite but minority view that saturated fats are good for you. But let me strongly emphasise that this is very much a minority view. Very few professional dieticians go along with this opinion.

Small amounts of coconut oil every now and then would not be a big problem for the average person. For example, I personally prefer butter changes begin to happen.

Second, the claim is false. Coconut oil smokes at 177°C, while olive oil does so around 180°C to 240°C (from extra virgin olive oil to the lower-quality pomace olive oil). It's not a huge difference. But if you want really high temperatures from your oil, go for safflower oil. It has a Smoke Point of 266°C.

(even though it is a saturated fat) over margarine, simply because it tastes better and has a long history of human consumption. But I have only small amounts, and only occasionally.

The current medical advice is to keep saturated fat intake to less than 10 per cent of total daily energy.

The Western diet is already high in saturated fats, so why add more? Indeed, Cancer Council Australia recommends reducing or avoiding a diet rich in saturated fats.

Another problem with coconut oil is that it's expensive. Coconut oil has no proven health benefits, yet it's about twice the cost of olive oil. But olive oil has proven health benefits. Why choose a lesser oil that is twice the price? (I guess expensive marketing counts for a lot!)

I like eating raw coconut occasionally, as part of a balanced diet – and for the sheer pleasure of it all. But based on the evidence, a real health nut would avoid coconut oil and coconut water.

17

OLDEST KNOWN COMPLAINT LETTER

MOST PEOPLE DO NOT DELIBERATELY MAKE MISTAKES, BUT THAT DOESN'T STOP THE SUFFERERS OF THOSE MISTAKES FROM WANTING TO COMPLAIN WHEN THEY HAPPEN.

And how long has this been going on? A *long* time.

The Oldest Known Complaint Letter – and let me emphasise, *written* letter – goes back 3750 years.

WRITTEN LANGUAGE 101

The oldest written languages were invented by civilisations that lived along rivers. It's not a coincidence – rivers give you easy access to food, water, clay for building and transport.

The Chinese developed alongside the Yellow River, the Indians along the Indus River, the Egyptians along the Nile River, and the Mesopotamians between the Tigris and Euphrates Rivers in the Middle East. Maybe living between two great rivers made it easier for the Mesopotamians to develop a civilisation.

The word "Mesopotamia" comes from the Ancient Greek – "meso" meaning "middle", and "potamia" meaning "(land) between the rivers". Mesopotamia covered almost all of present-day Iraq and Kuwait, along with smaller parts of Syria, Turkey and Iran.

Writing in Mesopotamia began with pictographic scripts, about 6000 years ago. Documented records of historical events involving early kings of the region survive from about 500 years later. During the Babylonian Empire, both men and women learned to read and write.

The long and complex history of Mesopotamia encompasses some 12,000 years. Many scholars consider it to be the Western Cradle of Civilisation. This history began with a Pre-Pottery Age. It continued through the Early, Middle and Late Bronze Ages to the Iron Age (and finished off with Classical Antiquity, then Late Antiquity around the 7th century AD. Mineral use from humans' native environment was so important to the development of society that scholars classify older societies as Stone Age, Bronze Age and Iron Age.)

MESOPOTAMIAN CULTURE

Mesopotamian society was advanced and complex.

The Mesopotamians had a mathematical system based on 60 (rather than on 10, as we have today). This had advantages – 60 could be divided into 2, 3, 5, 10, 12, 15, 20 and 30 (while 10 can be divided only by 2 and 5.) This is the historical reason why today we have 60 minutes in an hour, 24 hours in a day (two lots of 12) and 360 degrees in a circle. Following on from their mathematics, they developed some of the first known maps. Mathematics was also essential to their sophisticated astronomy.

Mesopotamian medicine also evolved over the millennia. The Babylonian medical text *Diagnostic Handbook* was written by Chief Scholar Esagil-kin-apli of Borsippa, during the reign of the Babylonian king Adad-apla-iddina (1069–1046 BC). It covers causes of diseases, lists of symptoms with detailed observations, rational diagnoses, physical examinations, therapies, prescriptions and prognoses (outcomes).

The Mesopotamians had agriculture. They invented irrigation, water storage and flood control to make best use of their semi-arid landscape, swamps and two rivers. They were among the first to make beer and wine.

They had armies – as well as philosophy and religion. At times, their temples were also banks, with large-scale loans and credit. They had monthly ceremonies (based on many factors) and music for both royalty and commoners. Songs were a way of passing on knowledge of historical events over decades and centuries. The Mesopotamians played organised games similar to modern boxing, wrestling, and rugby (with a wooden ball), as well as a kind of polo played on the shoulders of other men rather than on horses, and board games.

They developed many technologies such as glass- and lamp-making, textile weaving and metalworking. In fact, thanks to their metalworking, they were one of the first Bronze Age peoples in the world. Over the millennia, they worked their way through various metals – copper, bronze alloyed with both arsenic and tin, gold, and finally iron.

Thanks to its broad and complex society, trade was vigorous in Mesopotamia. That included trade in copper ingots, the subject of the oldest written complaint letter.

SMELTING

Smelting is the process of extracting a pure metal from its ore. Our ancestors knew of seven major metals – gold, copper, lead, silver, tin, mercury and iron.

The oldest evidence of smelting of any metal is a set of cast lead beads, in Turkey, about 6500 BC.

It usually isn't necessary to smelt gold, because gold is so unreactive that it is often found as pure lumps. Copper was also occasionally found as lumps of pure copper, but was mainly found as a mineral or compound (e.g. carbonate, sulfide, oxide, etc.). The other five metals were always found as minerals, and had to be "smelted".

Smelting uses both heat and a chemical reducing agent (commonly carbon). The ancients used quite sophisticated chemistry to get the raw metal.

The earliest copper smelting we know of dates back to 5500 BC, in Serbia. If you were smelting copper back then, the starting point was often the ore called "malachite" – copper carbonate hydroxide, or $Cu_2(CO_3)(OH)_2$. The only atoms the metal workers wanted were the copper (Cu) atoms – the rest (carbon, C, oxygen, O, and hydrogen, H) were useless to them.

In the first stage, the malachite was heated in several successive chemical reactions to temperatures ranging between 250°C and 350°C. This drove off carbon dioxide and water (CO_2 and H_2O). This left copper oxide (CuO). In the final stage, the copper oxide was heated with carbon monoxide (CO), which dragged the last oxygen atom off, leaving behind relatively pure copper.

OLD-SCHOOL COMPLAINT LETTER

The oldest known written complaint letter comes from the First Babylonian Dynasty, about 1750 BC (in the Middle Bronze Age). This was around when one of the most renowned kings of Mesopotamia, Hammurabi, had promulgated one of the earliest known set of laws – the Code of Hammurabi. (All societies need some kind of Rule of Law. Read *The Better Angels of Our Nature*, by Steven Pinker).

This complaint letter was written in cuneiform. The Latin word "cuneus" means "wedge", while "form" means "shape" – so "cuneiform" was a series of wedge-shaped marks on clay tablets. A blunt reed was used as a stylus to make the marks. Cuneiform originated about five and a half thousand years ago, and died out around 200 AD. Only in Roman times was it replaced by alphabetic writing, which had advantages so great that it led to the complete disappearance of cuneiform. Cuneiform was so thoroughly "lost" that it was deciphered only as recently as the mid-1800s.

In 1967, A. Leo Oppenheim published his book, *Letters from Mesopotamia: Official, Business and Private Letters on Clay Tablets from Two Millennia*. Rather than deal with epic texts, royal promulgations and official pronouncements, he specifically selected his 150 clay tablet translations to give a more intimate and diverse image of the Mesopotamian civilisation. He covered a period from around 2300 BC right up to 540 BC. The letters were mostly from Mesopotamia, but included some from as far away as Asia Minor, Cyprus and Egypt. Oppenheim includes a plea from a women with dependants to her sporadically generous brother, an entreaty from a pregnant slave girl to her master and, yes, various communications between merchants, manufacturers and traders.

The backstory begins back in 1750 BC. A copper merchant called Nanni sent a letter (yep, the Oldest Complaint Letter) to a smelter of copper called Ea-nasir. Copper was commonly traded in the Persian Gulf around that time. Nanni complained about the poor quality of the copper ingots his agent was being offered, the rude treatment given to his agent, and how he (Nanni) now had neither copper ingots nor his money.

Nanni wrote, "You have put ingots which were not good before my messenger and said, 'If you want to take them, take them. If you do not want to take them, go away'. What do you take me for that you treat somebody like me with such contempt? I have sent as messengers gentlemen, like ourselves, to collect the bag with my money (deposited with you), but you have treated me with contempt by sending them back to me empty-handed several times, and that through enemy territory.

... You alone treat my messenger with contempt! ...

You have withheld my money bag from me in enemy territory; it is now up to you to restore my money to me in full."

There you have it. A slice of life from nearly 4000 years ago – an angry merchant ripping strips off an unreliable supplier.

Now it turns out that a million or so cuneiform tablets have been excavated over the years, but only about 5 to 10 per cent have ever been translated. Why? Well, there's only a few hundred qualified cuneiformists on the whole planet. So, if you want to enter an interesting field that is not already overflowing, go and learn Mesopotamian cuneiform. At the same time, you would learn more about daily life several thousand years ago.

In today's world, if something goes wrong in a commercial or human transaction, you just bang off an email, and it arrives within a second.

But four thousand years ago, while still seething with anger, you had to first find a scribe who would spend time writing your complaint in cuneiform onto a moist lump of clay, and then fire it carefully in an oven. Once it had dried and cooled down, you then had to get a courier to personally deliver this fragile tablet very carefully over hundreds or thousands of kilometres – sometimes through enemy territory. "Customer-Feedback Technology" has changed enormously.

Remember *that* the next time you go to fire off an angry text, tweet or email ...

AN ANCIENT LETTER FROM A SISTER TO HER BROTHER

Tell my little Gimillum, whom my lord (the god Samas) keeps in good health; the *naditu*-woman Awat-Aja sends the following message:

May my Lord and my Lady (the goddess Aja) keep you for my sake in good health forever.

When I saw you recently, I was just as glad to see you as I was when (long ago) I entered the *gagum*-close and saw (for the first time) the face of my Lady (the goddess Aja). And you too, my brother, were as glad to see me as I to see you. You said: "I am going to stay for ten days." I was so pleased about it that I did not then report to you on my situation; I did not want to tell you here personally what I used to write to you about, before, from a distance. But you left suddenly and I was almost insane for three days. I did not touch food or even water.

You well know the amount of barley which I received before, and which you yourself had sent me. (If we continue) in this manner we will not wrong each other and I will not die of hunger with my household. Just send me the amount of barley which it was customary to send so that I can keep my household provided with food, that cold and hunger should not plague me (during the coming cold season).

PS: Have a heart, my dear Gimillum, let me not die of hunger. I was more pleased with you than I was ever with anybody else.

18

BREAKING THE SEAL

IF YOU'VE EVER HAD A LONG NIGHT OUT AT THE PUB, YOU WILL HAVE RUN ACROSS THE STRANGE PHENOMENON CALLED "BREAKING THE SEAL". YEP, WE'RE TALKING YELLOWISH BODILY FLUIDS.

Your bladder holds up well until around the third (or fifth) drink. Then you just have to "go" (and urinate) – you "break" the magical "seal". For the rest of the night, you have to keep popping back to your new best friend, the toilet.

Of course, there is no special "seal" keeping the urine inside your bladder – only your normal urinary sphincter.

Excessive urinary flow happens for two reasons – alcohol makes you generate more urine than usual, and you're drinking a lot more liquid than usual.

DRINK ONE, URINATE TWO

As far as water is concerned, your body is a dynamic powerhouse. In each 24-hour day, about 50,000 litres of water crosses the many, many membranes in your body – and practically all of it (yep, 50 tonnes) crosses right back again. The individual water molecules cross these membranes back-and-forth hundreds of times each second.

Each day, only a few litres of liquid actually leave your body – as urine, water vapour from your mouth, sweat and so on. They are replaced by water in the food you eat and in the liquids you drink. So under normal conditions, your body easily maintains its natural balance.

But alcohol interferes with the ability of your body to maintain this balance – and proper hydration. For every 200 millilitres of beer you drink, your kidneys generate 320 millilitres of urine. (See "Alcohol and Dehydration" in my 32nd book, *50 Shades of Grey Matter*.)

So when you guzzle down alcoholic drinks, you generate more urine than you would as compared to drinking the same amount of plain water.

DEHYDRATED, BUT LIGHT URINE?

People who work in environments where they might get dehydrated are taught how to monitor their state of hydration. They observe their own urine. This leads to a paradox.

If there's lots of urine and it's light in colour, they are probably well hydrated. But if they produce only a scanty urine flow, and if it's dark in colour, they may be on the way to dehydration.

So how can it possibly be that after "Breaking the Seal", a person simultaneously gets dehydrated *and* generates lots of light-coloured urine. What's going on?

Alcohol is a powerful drug. It has thrown their normal protective systems out of whack. A person (in their normal state) should generate less urine as they get dehydrated. Instead, because of the drug alcohol, they generate more.

PART 1: ANTI-DIURETIC (-URINATION) HORMONE

Dehydration is quite serious – you can die of it. So your body has several mechanisms to keep you well hydrated.

One of these involves a hormone that confusingly has two common names – Vasopressin, or Anti-Diuretic Hormone (ADH). "Diuretic" just means "related to urination". So the main job of ADH is to stop you from urinating.

Suppose you're lost in the desert and you run out of water. Your brain makes lots of ADH, and so you virtually stop making urine. That's a good survival tactic.

But suppose you drink alcohol. The awkward side effect of drinking alcohol is that your brain makes hardly any ADH, and so your kidneys start making lots of urine.

PART 2: VOLUME

But the other factor related to "Breaking the Seal" is simply Volume. If you're heading for a Big Night Out, you're probably not going to drink half a dozen big glasses of water, or milk. But you might easily drink half a dozen big glasses of beer.

So let's compare your bladder to a dried-out sponge.

Start dripping water on to this sponge, and for quite a while, it will just soak it up and store it. But eventually, the sponge will become totally saturated with water. Add just one more drop of water on the top of the sponge. Now a single drop of water will form on the bottom of the sponge. Add more drops, and a continuous trickle begins to form.

Your bladder has a capacity of around 600–800 millilitres. But of course, you need to "go" long before your bladder is completely full.

You throw down the first few beers pretty quickly. But remember, alcohol massively reduces your production of ADH. So for every 200 millilitres of beer you drink, your kidneys generate 320 millilitres of urine. By the time you get up to a litre of beer, you're getting a little dehydrated, and your kidneys will cut back a little – but overall, the volume of urine you generate is still greater than the volume of water contained in the beer.

So you do get slightly dehydrated, because you are urinating like crazy.

DRINK MORE WATER?

What if you drink lots of water to compensate for the extra volume of urine you generate? Your body will hang onto only about 30–50 per cent of the extra water – the rest just goes down the toilet bowl. You end up better off than having not drunk the extra water – but you'll still be dehydrated.

The moral is quite straightforward – you don't buy beer, you only rent it. And it's unlikely you'll ever regret *not* drinking too much.

19

WINE GLASS SHAPE AFFECTS FLAVOUR

WE HUMANS HAVE DRUNK ALCOHOL FOR THOUSANDS OF YEARS. THIS MEANS THAT WE'VE HAD PLENTY OF TIME TO MAKE MANY DIFFERENT SHAPES OF GLASSES TO DRINK ALCOHOL FROM. WE'VE GOT SHOT, BRANDY AND MARTINI GLASSES, AND EVEN SEVERAL DIFFERENT TYPES OF WINE GLASS.

Some rather recent and neat technology can show alcohol concentrations in the air in real time. These analyses seem to have confirmed that wine connoisseurs' claims could well be right – i.e. that the shape of the wine glass can affect the aroma and flavour.

SEEING IS SMELLING

Dr Kohji Mitsubayashi and colleagues at Tokyo Medical and Dental University developed the technology. It can directly image, in real time, ethanol vapour escaping from a glass. (There are many alcohols, but the one that we commonly drink is called "ethanol".)

They started with a cotton mesh. It was one millimetre thick, had tiny holes about one millimetre across, and measured 80 millimetres by 80 millimetres. They added a whole bunch of very specific chemicals, including one called "luminol". They then part-filled the test glasses with a Japanese red wine. As the ethanol rose upwards, it hit the cotton mesh. A bunch of chemical reactions started, and the luminol began to glow. By 50 seconds, the light output of the glowing cotton mesh had stabilised. They repeated the experiment with various glasses, and at different temperatures.

The researchers found that ethanol vapour appeared in a very characteristic ring shape. Alcohol concentrations were low towards the centre, but high in a doughnut-shaped ring just inside the rim of the wine glass.

Special cotton mesh

GLASS AND WINE

Because the alcohol concentration was lowest in the centre of that glass, that made it the best location to sample the delicate aromas of the wine. The lack of alcohol allowed the aromas to come forth and present themselves. The authors wrote, "The shape of the wine glass has a very sophisticated and functional design for tasting and enjoying the aroma of wine". On a similar but different theme, the shape of the drinking glass can affect how much you drink. (See "Drinking Glass Shapes Perception" in my 34th book, *Game of Knowns*).

Certainly, over the years, wine connoisseurs have maintained that the same wine can have very different bouquets and finishes depending on the wine's temperature and the shape of the glass it's served in. For example, it's been claimed that the tiny bumps present on the surface of lead crystal glass can bring forth a multitude of extra fragrances.

The wine scientist Régis Gougeon, from the University of Burgundy, France, admired the new technique, which enables a very simple and inexpensive, yet graphic, detection of ethanol. He said, "This work provides an unprecedented image of the claimed impact of glass geometry on the overall complex wine flavour perception, thus validating the search for optimum adequation between a glass and a wine." This new technique allows better matching of the wine to the glass in which you serve it.

TEMPERATURE AND ETHICS

But there are two things to note.

First, the effect happened at 13°C, and only in the wine glasses. The effect vanished at 17°C. It was most pronounced in a Pinot Noir wine glass.

Second, in accordance with all known Ethics Committee Requirements, no wine was wasted in this study...

20

MICROMORT

WE ALL KNOW THAT YOUR WEIGHT IS MEASURED IN THE UNIT CALLED KILOGRAMS, AND THAT YOUR HEIGHT IS MEASURED IN THE UNITS OF METRES OR CENTIMETRES. BUT ONLY A FEW PEOPLE HAVE HEARD OF THE UNIT CALLED A "MICROMORT" – AND THEY MOSTLY WORK IN INSURANCE.

DEFINITION

The micromort measures risk. If you jump off a cliff, your chance of sudden death is one mort – you will almost certainly die.

(By the way, "mort" is related to "mortality", or death.) But most of our daily activities are much less dangerous than jumping off a cliff, so we have devised a much smaller and more useful Unit called a "micromort" – or one millionth of a mort.

ONE MICROMORT

There's a whole bunch of different activities that will expose you to one micromort.

They include drinking half a litre of wine, smoking 1.4 cigarettes, spending one hour in a coal mine, living for two days in New York or Boston – air pollution is the cause – or eating 1000 bananas. (Yep, as seen in "Slippery Banana Peels" on page 11, bananas are very slightly radioactive, thanks to their potassium-40 content. So are brazil nuts. Brazil nut trees have an extensive underground root system, which very effectively absorb any radium in the soil.)

Other food-related activities that expose you to one micromort of risk include eating 100 charcoal-grilled steaks (cancer from the benzopyrene from the cooking process) and eating 40 tablespoons of peanut butter (liver cancer from a fungus that makes aflatoxin).

INVENTOR OF THE MICROMORT

Ronald A. Howard was made a Professor at Stanford University in 1965 – and he's still there. He helped found the field of "Decision Analysis". Decisions are involved in every part of life, from investment planning, hurricane seeding and choice of a partner, to road design and nuclear waste disposal. Making a decision involves certainties, uncertainties, emotions, risks, etc.

Professor Howard introduced the concept of the micromort in 1980.

SEVERAL MICROMORTS

Each skydiving jump, or hang-gliding flight, exposes you to eight micromorts.

In other words you have an eight in one million chance of sudden death each time you skydive or hang-glide. You get seven micromorts for each marathon you run, and five micromorts for each bout of scuba diving.

Base-jumping is more dangerous – each event exposes you to 430 micromorts. And trying to climb Mt Everest gives you 40,000 micromorts in one ascent – that's a four per cent chance of death.

TRAVEL MICROMORTS

Getting out of bed is risky (but so is staying there). Travelling from here-to-there exposes you to the risk of sudden death.

When I was studying obstetrics and gynaecology, my professors told me the most dangerous journey you ever take in your whole life (in terms of injuries and death per kilometre) is the 10-centimetre journey down your mother's birth canal. And this truism is partly backed up by the risk assessors, who have figured that on your first day of life, your chances of dying are 430 per million (430 micromorts).With transport, to get a one-in-a-million chance of dying (one micromort), you need to travel 11 kilometres by motorbike, 32 kilometres by bicycle or 480 kilometres by car. Safest of all is the train at 9700 kilometres.

The average multi-lane divided expressway (1600 kilometres per micromort) is safer than rural two-lane blacktop (250 kilometres per micromort). The most dangerous road in the United Kingdom is the narrow, winding A18 that takes you through woods and farmland in Lincolnshire. Driving the round trip between Laceby and Ludborough (just 32 kilometres) on the A18's blind corners and hairpin bends exposes you to a full micromort.

Currently, the satellite navigation in your car will give you the option of the quickest or cheapest paths to your destination. Maybe it should also give you the safest route? SatNav or SafeNav?

21

GREAT BARRIER REEF

THE GREAT BARRIER REEF HAS LOST HALF OF ITS CORAL IN THE 27 YEARS BETWEEN 1985 AND 2012. IF THE TREND (WHICH IS ACTUALLY INCREASING IN PACE) CONTINUES, IT WILL LOSE HALF OF WHAT IS LEFT BY THE YEAR 2022.

AROUND 550 MILLION YEARS AGO ...

Evolution produced the first animals to have hard shells. They were *Cloudinia* – tiny, filter-feeding animals. Within two million years, they had built the oldest known coral reef in what is now called Namibia. Back then, this reef on a shallow equatorial seabed was about 300 metres thick, and about seven kilometres long.

Cloudinia attached themselves to fixed surfaces, and to each other, by extruding a "cement" based on calcium carbonate. Making a "reef" had many advantages, including protection from predators. The reef also created currents that were rich in nutrients.

What a shame for something so dazzlingly beautiful. It's the world's biggest single structure built by living creatures. You can see it from space. But it needs to be loved ...

CORAL REEF 101

The Great Barrier Reef stretches some 2300 kilometres along the coast of Queensland, with over 2900 individual reefs and some 900 islands.

Its surface area (350,000 square kilometres) is about one and a half times that of the state of Victoria. The Great Barrier Reef is located between 16 and 160 kilometres offshore. It's home to over 1600 species of fish, 133 species of sharks and stingrays, and over 30 species of whales and dolphins.

Back in 1842, Charles Darwin – Mr Evolution himself – wondered about coral reefs, thanks to his insatiable curiosity about everything. (The British Navy was also very interested, because it was constantly losing ships to reefs.) How did they form? With incredibly productivity, they flourished like an oasis in the desert – yet they grew in tropical waters that contained very few nutrients and fertilisers. How could they do this?

DRILL-AND-SPILL

In the 1960s and 1970s, the Queensland State Government tried to industrialise the reef. The Queensland Premier, Joh Bjelke-Petersen, actually invested in oil companies that he licensed to drill in the reef.

According to *The Guardian*, "One of his ministers even claimed that any oil spill would actually provide nutritious food for marine life, rather than kill it off."

We now know that coral reefs form an incredibly efficient food web and ecosystem that extracts a huge amount of energy from the environment. In a year, they can "produce" 35 tonnes of fish for each square kilometre.

Coral reefs cover less than one tenth of one per cent of the area of the world's oceans – but they support more than 25 per cent of all creatures that live in the oceans. They are the rainforests of the seas – our underwater Amazon. Unfortunately, they're in trouble.

On a global scale, about 50 per cent of the world's coral reefs have died in the past 50 years. Most of the rest are at risk. All this has been caused by human activities. The situation differs from place to place around the world – it's worse in Southeast Asia, and slightly better elsewhere. But no coral today is free from the effects of human activities.

Why on Earth would we destroy such productive sites?

CORAL POLYPS:
LARVAE TO SEXY ROCK

The Great Barrier Reef is made of, and was built by, trillions of tiny animals known as "Coral Polyps". There are some 400 different species

Great Barrier Reef   < 149

of Coral Polyps, both hard and soft, on the Great Barrier Reef.

Back around 1020 AD, the Muslim polymath Al-Biruni was one of the first to correctly classify corals as animals. He argued for this classification even though coral never changed their location – because they responded when he touched them.

In the Great Barrier Reef, coral gets "born" during the week after the Full Moon in October (inner reef) or November/December (outer reef). All the coral usually release their "eggs" and "sperm" on the same night. They are timed by special "love molecules" (cryptochromes) that detect the intensity of the moonlight falling on them. It's said to be a spectacular sight when they spawn. (I've never seen this, but I want to.) At this time, the difference between high and low tides is very low, and this makes it easier for the spawned eggs and spawn to "get together" and not get lost. (But besides this regular "sexual" reproduction, they also do "asexual" reproduction. They do "it" in many different ways – beside sexual and asexual reproduction, some can become hermaphrodites).

Once fertilised, the Coral Polyps swim around as soft-bodied larvae. They are attracted to light (so they head for the surface, rather than deep water) and sound (so they head for a reef, rather than for open waters). Then, if lucky, they will land on something hard, and take algae into their bodies. Algae are essential partners in a remarkable long-term relationship. The time between spawning and settling down is usually two to three days, but occasionally can reach two months.

Each Coral Polyp is a cylindrical animal usually only a few millimetres in diameter and a few centimetres long.

They sit on rock, or on the remains of previous coral, and excrete a carbonate exoskeleton at their base. At the other end they have a set of tentacles surrounding a central mouth. This lets them get at their food. Their "mouth" is also how the Coral Polyps eject their undigested waste. (In humans, our mouth is also a major "excretory orifice". Read "Fat: Where Does It Go?" on page 245.) Then the Coral Polyps grow into adults.

CORAL POLYP AND BESTIE NEED EACH OTHER

Most Coral Polyps cannot survive by themselves. They are relatively simple multicellular creatures. They don't have respiratory or circulatory organs.

Coral Polyps live in a very cosy and mutually beneficial relationship with tiny single-celled photosynthetic algae. These algae are called "*Symbiodinium*", and belong to a larger group colloquially called "zooxanthellae".

The Coral Polyps give these algae carbon dioxide, ammonium, nitrogen, phosphorus, sulfur and other "waste" products. These chemicals are scarce in the relatively barren waters that surround most coral reefs.

In return, the *Symbiodinium* give the polyps essential carbohydrates such as glucose, as well as glycerol, amino acids and other nutrients so they can grow. Indeed, the *Symbiodinium* provide more than 90 per cent of the metabolic needs of the Coral Polyps inside which they live.

The relationship is so close that the *Symbiodinium* actually live inside the flesh of the host Coral Polyp. They reside in a thin layer of animal flesh that covers the hard skeleton of calcium carbonate. There are about a million of them in each cubic centimetre of coral animal flesh.

Symbiodinium are really quite weird – unlike any other life form on our planet. They have 100 times more DNA than we humans have – and we have no idea why. And while all other known life on the planet has four different nucleotides (or base pairs) in their DNA Ladder of Life, *Symbiodinium* have five! They do photosynthesis – but by a completely different process from all other light-harvesting organisms. (If you want to get really speculative, they are almost like a life form from another planet.)

They are also one of the sources of colour of coral, thanks to their inherent brown-yellow colouration.

CORAL CURRENTS

At first glance, Coral Polyps seem to be totally at the mercy of their environment. They can't move from where they settled down. And their food is in the water around them.

They do get a reasonable flow of water over their bodies when the tide is flowing rapidly. But a few times each day, there's not much flow – when the tide is on the turn. And what about those Coral Polyps

COLOUR OF CORAL

Coral get their colour from two main sources.

Most of their colour comes from the golden-brownish pigments in the *Symbiodinium*.

A smaller part of their colour range (such as red, green and blue) comes from coloured proteins that the Coral Polyp itself makes and stores. These proteins, known as "pocilloporins", will glow in the dark when viewed under an ultraviolet blacklight.

So when coral lose their *Symbiodinium*, they might turn white (the natural colour of calcium carbonate, or limestone) if they don't have many of their own inherent coloured proteins. But if they do have these coloured proteins, they will fade to a pastel shade. As the brown colour fades, the iridescent colours of the pocilloporins become even more vibrant. This is often very beautiful – but tragic. We are actually watching an animal die in such a colourful way.

Without the brown colour, corals are said to be "bleached". This condition can be fatal.

who have been unlucky enough to land in a quiet lagoon, or inside a densely branched section of a reef, with tiny currents?

How do they survive?

It's recently been discovered that Coral Polyps make their own water currents. They wave their cilia (little hairs) and generate swirling currents that keep a thin layer of water moving parallel to their surface, like a mini conveyor belt.

This moving current, up to two millimetres thick, simultaneously brings in foody goodness and takes away waste products.

CORAL BLEACHING: MOSTLY HEAT DEATH

There are many causes of Coral Bleaching.

Overwhelmingly, overheating from climate change is the main cause. Unfortunately, most corals live in shallow waters, and are already near their maximum survival temperature.

Worldwide Coral Bleaching began happening in the late 1970s, when atmospheric carbon dioxide levels reached around 320 parts per million. Since then we have seen Bleaching Events become distressingly more frequent – and more severe.

Between 1980 and 2010, there were some 60 major episodes, globally. On the Great Barrier Reef, there have been seven Coral Bleaching events in the summers of 1980, 1982, 1992, 1994, 1998 (the Big El Niño Year), 2002 and 2006. With Global Warming, Coral Bleaching is expected to become far more frequent – in some locations, it could become an annual occurrence.

But there are other causes of Coral Bleaching (sometimes related indirectly or directly to Climate Change). They include changes in seawater chemistry – ocean acidification, pollution from farm run-off into rivers, release of nitrate-laden waters by mining companies into reefs (such as happened to the Great Barrier Reef), etc. – sediment, ultraviolet light, disease, loss of coastal wetlands, cyanide fishing, sunscreen from tourists snorkelling or swimming, etc. But none of them match the devastation that heat stress has caused to coral reefs over thousands of square kilometres of ocean territory.

RUINING THE REEFS

A large number of the coral reefs in the world are being killed by unsustainable fishing practices. These include using dynamite and cyanide to stun fish, which then float to the surface and are easily collected.

Overfishing has nasty (and sometimes unexpected) flow-on consequences. Consider just one case – herbivore fish.

When the plant-eating fish (herbivores) are removed, seaweed flourishes. These herbivores are essentially the Gardeners of the Reef. (Just imagine what would happen to your vegetable patch if you ignored it for a year, or what would happen to your local Botanical Gardens if they suddenly fired all the gardeners.) Losing the herbivore fish leads to coral being overwhelmed by seaweeds. It also leads to fewer carnivorous fish and crustaceans (guard-crabs) that eat the herbivore fish. You need meat-eating predators in an ecosystem. There is evidence that the outbreaks of the coral-munching Crown of Thorns starfish are related to the loss of their predators.

Separately, there's other evidence that too much fertiliser favours the survival of the baby starfish. Rivers pick up fertilisers and pesticides from farms, as well as sewage. Between them, the sewage and fertiliser increase the growth of algae and seaweed. Rivers also carry huge amounts of sediment, coming from areas of deforestation. This sediment physically smothers the coral.

And without Gardeners, the whole ecosystem gets out of hand.

CORAL BLEACHING: THE PATHOPHYSIOLOGY

Remember that the Coral Polyps and *Symbiodinium* are in a close symbiotic relationship.

When they are stressed, *Symbiodinium* can lose much of their photosynthetic pigments. This means less foody goodness for their hosts, the Coral Polyps.

When Coral Polyps are under stress, they expel their essential symbiotic friends (the *Symbiodinium*) from their body. The Coral Polyps lose their rainbow colouration and turn white – so this event is called "Coral Bleaching".

This action is a desperate last-ditch survival effort. Temporarily, they don't have to feed their little friends. But unless they take up some *Symbiodinium* within a few months, the Coral Polyps are doomed and will die. A lush ecosystem turns into a barren and bleached landscape.

SUMMARY OF CORAL DEATH

Unfortunately, recent research has shown the Great Barrier Reef lost half of its coral in a tiny 27-year window.

About 10 per cent of that was directly due to Coral Bleaching. About 48 per cent was caused by storm damage, and about 42 per cent was due to attack by the infamous Crown of Thorns Starfish.

These are all related.

We have measured that Global Warming has raised the temperature (and energy levels) of both the atmosphere and the oceans. As a result, there's more energy in storms. We would expect more storm damage, and we're seeing it.

As the environment along the length of the Great Barrier Reef has changed, the Crown of Thorns Starfish has increased in numbers. These starfish sit on top of Coral Polyps, turn their stomach inside out, and then liquefy and absorb the flesh of the coral.

We also know that the Crown of Thorns Starfish prosper after floods.

Their current (2015) outbreak on the Great Barrier Reef stems from the enormous Queensland floods of 2010 and 2011. Climate Change is expected to create increased rainfall and more floods in between massive droughts – so it's also implicated in the outbreaks of the dreaded starfish.

CORAL CALCIFICATION

In a further blow, Global Warming changes the chemistry of the oceans, by stealing the chemicals – particularly calcium carbonate – needed to build reefs. (See the story "Ocean Acidification" in my 34th book, *Game of Knowns*.)

Ocean pH has dropped by 0.1 since the Industrial Revolution. Unfortunately, most models predict a further drop of 0.3 or 0.4 by the end of the 21st century.

On a good day, Coral Polyps can lay down 10 grams of calcium carbonate for each square metre of reef. They have to do this continually to keep ahead of erosion caused by waves and other forces. Their production is highest at midday, and lowest at night.

However, because the reefs are platforms made from calcium carbonate, they are very sensitive to changes in the chemistry of the oceans.

One major study looked at 328 coral colonies from 69 reefs covering most of the 2000-plus kilometre length of the Great Barrier Reef. These corals ranged in age between 10 and 426 years. The study showed that the rates of calcification increased by 5.4 per cent between 1900 and 1970.

But then reef calcification went the wrong way between 1990 and 2005 – the rates dropped by 14.2 per cent. This decline was seen in both onshore and offshore reefs. So this is a global ocean problem, not just one of declining coastal water quality in the Great Barrier Reef. For example, Bermuda brain corals have suffered a 25 per cent decline in growth and calcification since 1959.

The drop in coral reef calcification is a problem that is here right now. It's not a problem that might happen in the future.

And here's the kicker. The Geologists and Oceanographers tell us that it will take at least 10,000 years for the ocean to repair its chemistry. It will do this via the addition of alkaline material to the ocean from continental weathering. Just think, we are making changes that at least 300 generations of humans will have to live with!

FUTURE PROSPECTS?

According to Professor Ove Hoegh-Guldberg from the Centre for Marine Studies at the University of Queensland, if we don't do something about the threats to coral reefs, they "will become rapidly eroding rubble banks such as those seen in some inshore regions of the Great Barrier Reef, where dense populations of corals have vanished over the past 50 to 100 years".

We need our coral reefs. For example, coral reefs are a major source of protein for over one billion people in Asia. But elsewhere, the Caribbean has lost 80 per cent of its coral over the last half-century.

The global net economic value of coral reefs is variously estimated at between $30 and $375 billion per year – depending on how widely the Economists cast their net. In Australia alone, the income from international tourism due to the Great Barrier Reef is around $5–7 billion each year. The two million people who visit the Great Barrier Reef annually are supported by people in some 64,000 jobs.

Reefs also provide coastal protection, and huge numbers of bioactive chemicals for our pharmaceutical industry.

But the coral reefs are changing.

The best characteristics that an animal should possess to deal with a rapidly changing environment are a short life cycle and sexual reproduction. Sexual reproduction means the babies can be different from the parents, and so have a better chance of dealing with changes.

Unfortunately, the Coral Polyp has exactly the wrong features. It has a very long life cycle – five to one hundred years. Also, it often reproduces asexually by cloning, so that the babies have exactly the same characteristics as the parents.

So if the Coral Polyps can't change their ways in a hurry, maybe we have to – so that the Great Barrier Reef doesn't become the Mediocre Barrier Reef.

CORAL CAN SOMETIMES RECOVER

Back in 1998 (the most significant El Niño year so far on record) there was a worldwide mass Coral Bleaching Event. By September 1998, thanks to the temporary ocean heating, nearly one-sixth of the world's coral had bleached. This remains the largest Mass Bleaching Event on record.

Dr Nicholas Graham from James Cook University in Townsville, Queensland looked at 21 specific coral reefs around the Seychelles.

He followed the progress of these 21 reefs in the Seychelles for 17 years. He recently wrote up his findings. It turns out that after a long delay, slightly over half of the damaged coral reefs began to recover.

Back in 1998, these 21 reefs in the Seychelles had suffered much more than the worldwide average of Coral Bleaching. Early on, about 90 per cent of the coral on these reefs had gone. This coral didn't recover significantly over the next decade. Nine of the 21 reefs never recovered.

Beginning in 2005, a remarkable recovery occurred on the other 12 reefs. On these 12 reefs (out of the initial 21 that had all lost about 90% of their coral) the coral began to return. Before the 1998 El Niño bleaching, hard coral had covered 28 per cent of these reefs. Thanks to the El Niño, it dropped to around 0-10 per cent. By 2010, it had come back to 23 per cent. These 12 reefs are now on a path to full recovery.

Dr Graham identified five factors critical to survival. The reefs that recovered were deeper, had many nooks and crannies, were located in less polluted water, and had both young coral and lots of plant-eating fish. (Yes, seaweed and coral compete for the same pieces of real estate. So if the fish eat the seaweed, there's more real estate left for the coral.)

Unfortunately, at the time of writing, it seems that another El Niño looms. Those reefs Dr Graham studied may well get clobbered once again back to around zero coral. How many times can coral recover? We are doing the experiment to find out now.

22

ANTI-OXIDANTS & SNAKE OIL

ANTIOXIDANTS ARE BIG NEWS. WHEN YOU GO INTO HEALTH FOOD SHOPS, THERE ARE ROWS OF ANTIOXIDANT SUPPLEMENTS FOR SALE.

However, the claims for antioxidant attributes are a mix of the Half-Truth, and Marketing Exaggeration. It seems that supplemental antioxidants are the 21st century version of Snake Oil – that expensive "cure-all medicine" that is 100 per cent pure rip-off.

It turns out that – depending on the circumstances – antioxidants can be good for you. Or they can be bad for you. It's the same for oxidants. Our "popular" concepts of oxidants and antioxidants seem to be deeply flawed.

WHAT "THEY" TELL YOU

By the way, antioxidants don't come only in tablets, capsules or expensive drinks. They also occur naturally in food. But "Food" doesn't sound as exciting as "Superfood".

So food suppliers sell what they claim are antioxidant-rich Superfoods. These Superfoods are claimed to do everything from preventing cancer and chronic disease, to slowing ageing. (And improving your handwriting? No, just kidding.) They include green tea, goji berries, chia seeds and even red wine and dark chocolate. (Read about green tea in my 36th book, *House of Karls*. The evidence for health benefits from green tea is weak, but it does taste nice. Unfortunately, sometimes it contains lead.) You might not have studied Biochemistry or Dietetics, but you are probably familiar with the basic claims about antioxidants.

Plain and simple, "oxidants" (whatever they might be – after all, who knows chemistry?) are bad. Obviously then, antioxidants – presumably the "opposite" of oxidants – must be good.

And if a little antioxidant is good, surely more must be better.

There you have it – the antioxidant claim in a nutshell. But the antioxidant story isn't that simple.

FREE RADICAL DAMAGE

There's no doubt that the Free Radicals (oxidants) your body makes are associated with bodily harm.

They are implicated in some of the degenerative changes of ageing, as well as damage and even mutations to DNA. They're also involved in atherosclerosis (so-called "hardening of the arteries"), dementia, and vision loss via macular degeneration and cataracts.

It turns out that your own body also makes antioxidants – partly to counter Free Radicals. They act to "mop" them up.

But as you shall shortly see, Free Radicals can have a good side.

OXIDANT CHEMISTRY

First, what is an oxidant? There are a few different ways to answer this.

Oxidants are chemicals that transfer atoms (usually oxygen) to another chemical. Another way of putting it is that oxidants are chemicals that remove an electron from another chemical. That's why oxidants are also known as "Free Radicals" – extremely reactive chemicals.

To summarise, oxidants are pretty reactive chemicals. They are the reason why iron rusts, or a cut apple turns brown.

Oxidants come from many sources. So one way to look at oxidants is to see them as entirely natural chemicals that can be produced when the cells in your body turn food into energy. But oxidants can also be made by sunlight in our skin and eyes, cigarette smoke, alcohol, air pollution and even exercise – yes, exercise. And further, plants – which are often thought of as sources of only antioxidants – can deliver oxidants into your body.

It turns out that oxidants are a twin-edged sword – they're essential for life, but they can also cause damage. (And yes, the same holds for antioxidants, and water, and food, and sunlight.)

STRESS – GOOD OR BAD?

Dr Toren Finkel is Chief of the Center for Molecular Medicine at the National Heart, Lung and Blood Institute in Bethesda, Maryland. He says that the simple concept that oxidants are bad is "oversimplified to the point of probably being wrong. Oxidants may be a primordial messenger of stress in our cells, and a little bit of stress, it turns out, may be good for us."

ANTIOXIDANT CHEMISTRY

Antioxidants are molecules that tend to slow or block oxidation. Antioxidants tend to fall into two main classes – they dissolve either in water, or in fat.

Fat-soluble antioxidants include Vitamin A and beta-carotene (which you get from the yellow, orange and reddish fruits and vegetables), as well as Vitamin E and Co-enzyme Q. These fat-soluble antioxidants often protect cell membranes.

The water-soluble antioxidants include Vitamin C and glutathione. They react mainly with oxidants that exist in the water inside your cells, and in your blood.

Where do they come from? Plants and us.

Most of us know that plants make antioxidants. They use them for many purposes – one of which is, almost certainly, related to self-preservation. After all, plants don't have arms or legs – they are sitting targets. They have to survive attacks from tiny pathogens and large grazing animals, as well as heat and cold, drought and flood or too much ultraviolet radiation from sunlight.

But here's a surprise. Our bodies will, under the right circumstances, make their own antioxidants. (That's right, our bodies can make both oxidants and antioxidants.)

Dr David Sinclair from Harvard has taken this concept further. He and his colleague, Dr Konrad Howitz, have developed the concept of "xenohormesis", which they interpret as "benefitting from the stress of others". They see plants as manufacturing various antioxidants when they are stressed. Their theory is that you eat the food, absorb the relevant oxidants/antioxidants into your body, "sense" the harsh environmental conditions that the plant experienced, respond by making your own relevant chemicals – and enjoy better health as a result.

Is xenohormesis real? Maybe. We don't know, yet.

ANTIOXIDANTS CAN BE BAD

I'll give you two examples where antioxidants are good for you as part of a regular diet – but only when you ingest them as "regular" food, not when you take them as supplements. One involves exercise, the other cancer.

We know that physical exercise helps build muscle, and improves your life expectancy. Exercise can also improve your risks with regard to Type II Diabetes and Insulin Resistance.

Exercise forces your muscles to make lots of Free Radicals (oxidants). In response to your muscles producing oxidants, other cells in your body produce antioxidants.

But in 2009, it was discovered that if you exercise and take antioxidant supplements, suddenly you lose many of the health-promoting effects of exercise. There are three major losses.

According to this study, expected increases in antioxidant levels (such as glutathione) don't occur, muscles don't get bigger, and you don't get protection from diabetes. In other words, antioxidants from a bottle are bad for you. They are mopping up natural oxidants – and stopping them from doing their Good Work. (Let me emphasise, this is not the case when you get your antioxidants by eating food in regular quantities.)

We found a similar disturbing result with lung cancer.

We all know that smoking cigarettes can cause lung cancer. (OK, most of us. About 4 per cent of Americans deny this cause-and-effect link between tobacco and cancer.) If you have a large enough group of people with lung cancer, you can get a pretty accurate idea of how long they will live – on average.

If you then give these people with lung cancer a diet rich in natural antioxidants (from foods), it turns out that they will live a bit longer than expected.

It seems reasonable to expect that if a little bit of antioxidants is good, then lots of antioxidants will be better. So the researchers tried giving extra supplemental antioxidants – from a bottle, not food. The results were astonishing. People didn't live longer. In fact, they died sooner. The additional antioxidants shortened their life expectancy! (Read more in my 24th book, *Disinformation*, in the story "Vitamin OD"). We've since found the same effect in mice with lung cancer.

This is still early days, and we don't have the full story yet. But let me tell you what we do have.

OXIDANTS CAN BE GOOD

So the simplistic claim that antioxidants are always good is incorrect. What about the other half of the claim? Are oxidants always bad?

Sometimes oxidants can be bad for you. But they can also be good for you – regardless if they are made by your body, or plants.

For example, as I've already mentioned, when we humans exercise, we manufacture oxidants.

Is this bad? No.

Our natural oxidants trigger our cells to make our own natural antioxidants in response, which then give us the benefits of exercise. Not only do our natural antioxidants make our muscles stronger, they have unexpected side effects – such as to protect us from air pollution. However, we need all these steps in order: first we exercise to make our natural oxidants, then we get the benefits.

Next, consider plant oxidants and human health. Broccoli and other cruciferous vegetables, such as Brussels sprouts, make several oxidant chemicals. Of course, one of their chemical functions is protection of the plant.

But these chemicals can also have many health-promoting functions.

One of these chemicals is sulforaphane – a mild oxidant. It gives a mild sulfurous smell to Brussels sprouts and its relatives. When we humans eat sulforaphane, it triggers our bodies to produce a protein called "Nrf2". In turn, Nrf2 activates over 200 genes in our body. These genes then stimulate our cells to make a wide range of chemicals, including antioxidants, enzymes (to metabolise toxins), other proteins (to flush out heavy metals) and so-called "factors" (to suppress tumours).

ANTIOXIDANTS – FOOD VS PILLS

Let's go back to the antioxidants in natural foods. (Just ignore Superfoods for now.)

There are many natural antioxidants. They include flavonoids from tea, coffee and berries. There are also phytoestrogens from soy beans and peanuts, minerals such as selenium and manganese from whole grains, nuts, seafood and lean meats, and even resveratrol from red wine and dark-skinned grapes.

It seems that they work inside our bodies in the same way as our natural antioxidants that we trigger with exercise. They stress us – but in a good way.

And while I'm talking about natural food, don't forget all the other good stuff in whole foods – fibre and nutrition, and all the other "foody" goodnesses we haven't discovered yet. This bioactive package almost certainly works hand-in-hand with the antioxidants from natural foods. (Read more in my 31st book, *Brain Food,* especially "The Food Industry Under the Microscope").

But what about the supplemental antioxidants from a bottle? It seems that they actually prevent your body from making its own natural antioxidants. Even though the supplements do provide antioxidants, they don't balance the loss of your own native antioxidants.

A 2012 meta-study looked at 200,000 healthy people and 81,000 people with various diseases. The study found absolutely no benefits from antioxidant supplements. In fact, they found the opposite. The death rate definitely increased in those taking beta-carotene, and possibly in those taking Vitamins A and E. As a result, a US Government body, the US Preventative Services Task Force, specifically advised healthy citizens not to take Vitamin E or beta-carotene supplements.

Antioxidant supplements are at best useless, and at worst harmful. At the moment, the supplemental antioxidant industry generates take over US$23 billion a year in the USA (that's just antioxidants – not the total Natural Health Market). But these health fads come and go.

Nature is hard to repackage – especially if you're trying to get it into a pill.

ROCKSTAR PLANTS

Plants defend themselves, and counterattack, by producing chemicals called "phytochemicals". They also use them to help reproduction.

In terms of defence, tobacco plants make nicotine to fight off insects that graze on them. Beans fight off insects with different chemicals – lectins. Garlic makes allicin – an antifungal agent.

As an example of reproduction, consider the coffee plant which makes caffeine. You might not realise this, but flowering plants (such as coffee plants) have to compete for visiting rights by their pollinators, such as bees. Attractive plants get more visits. It appears that once a bee has tasted caffeine, its brain gets chemically "adjusted" so that in future it prefers to visit coffee plants. Addiction? Maybe. But either way, coffee plants are into Sex and Drugs.

FOX TELEVISION'S NEWS IS QUITE UNABASHEDLY OPEN ABOUT ITS PREJUDICES. FOR EXAMPLE, IT HAS A PHILOSOPHY THAT GOVERNMENT SHOULD NOT INTERFERE WITH PRIVATE ENTERPRISE – AND IN FACT, THAT PRIVATE ENTERPRISE IS BETTER IN PRACTICALLY EVERY WAY THAN GOVERNMENT. CONSIDER THIS EXAMPLE.

The US government no longer flies cargo up to the International Space Station (ISS). Instead, NASA has subcontracted two private companies, Orbital Sciences Corporation and SpaceX, to deliver cargo to the ISS in Low Earth Orbit (some 400 kilometres above the ground).

On 28 October 2014, Orbital Sciences was on target for a launch of its Antares rocket. Fox News had already announced, "Private Company Launches Antares Rocket to ISS."

Unfortunately, during the launch, one of the two Russian-made Aerojet AJ-26 rocket engines exploded. Fox News then announced, "Unmanned NASA Cargo Rocket Explodes on Launchpad."

Notice the swapping of "private company" for "NASA".

There were two mistakes in that second headline.

First, the rocket did not explode on the launchpad. It left the launchpad and, while it was in the air, exploded – some 15 seconds after take-off.

Second, it was not a NASA (government) cargo rocket. It was a private-enterprise rocket launched by Orbital Sciences.

To quote *New Scientist*, Fox News "clearly know who to praise and who to blame."

NEWSFLASH!

The company Orbital Sciences holds a US$1.9 billion contract with NASA to make eight cargo runs to the ISS. Its Antares rocket stands some 40 metres tall, and weighs 240 tonnes at take-off. It can carry up to 2.7 tonnes of payload to the ISS. On the launchpad, the combination of rocket and payload costs about US$200 million.

The Russians have a long history of making simple, rugged and reliable rocket engines. The AJ-26 is fuelled by liquid oxygen and kerosene. Its heritage goes back to the 1960s, when it began life as the Kuznetsov Design Bureau NK-33.

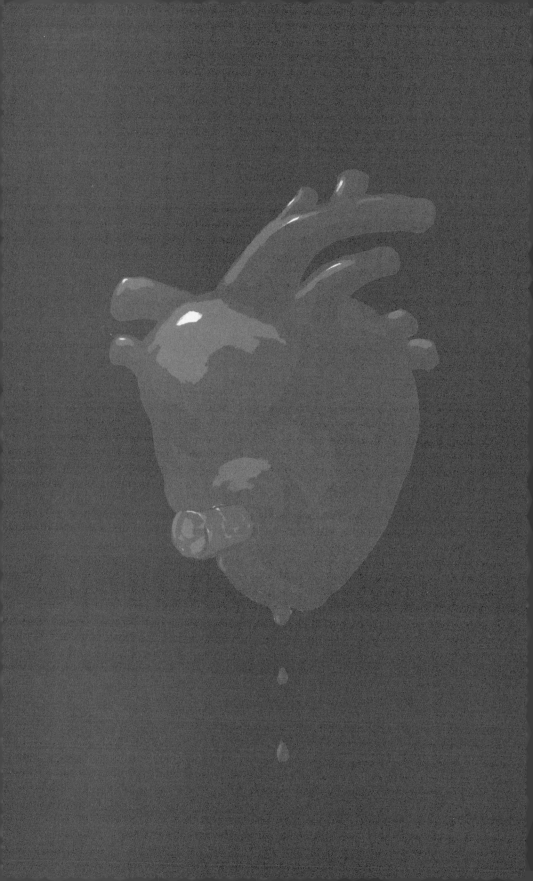

24

HOLE HEARTED

IT SEEMS THAT EXERCISE CAN REDUCE THE RISK OF YOUR BABY HAVING CONGENITAL HEART DISEASE – THAT IS, IF YOU'RE A MOUSE . . . TO BE EVEN MORE SPECIFIC, IT CAN REDUCE THE RISK TO YOUR BABY IF YOU'RE A MOUSE BRED TO HAVE A VERY HIGH RISK OF HAVING A HOLE BETWEEN TWO OF THE CHAMBERS IN THE HEART.

HOLE IN HEART 101

In humans, the heart is one of the first organs to develop in the embryo, usually within 10 weeks of conception.

Not surprisingly, it's quite complicated to turn the initial tube into a fully functioning pump with four chambers – two low-pressure chambers (the atria) and two high-pressure chambers (the ventricles). The wall that develops between the two atria, and the two ventricles, is called a "septum". This labyrinthine developmental process is regulated (or controlled) by the *Nkx2–5* gene.

We know that in humans, congenital heart disease is very common – about one per cent of all human births. A common congenital heart defect is the "classic" Hole In The Heart. This is either a hole between the right and left atria (Atrial Septal Defect, or ASD) or between the right and left ventricles (Ventricular Septal Defect, or VSD). About a tenth of these Holes In The Heart need surgery after birth – the rest heal themselves.

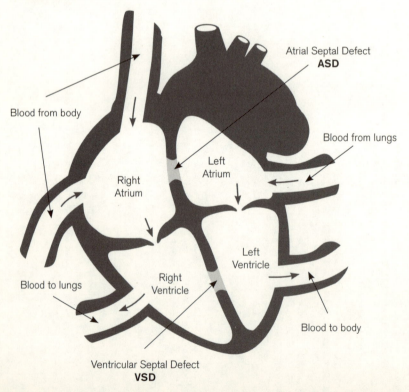

We would love to know anything that could possibly lower the risk of heart defects in human babies. This would reduce both the number of defects and the need for surgery. The mother provides the environment for the growing foetus, so looking at her risk factors is a good starting point. Maternal risk factors for congenital heart disease include certain infections in the mother, her genetics, maternal diabetes, environmental factors such as toxins, and finally, the age of the mother.

CONGENITAL?

In Medical Jargon, "congenital" simply means "present at birth". If you go back to the Latin roots, "con" means "with", while "genital" means "birth".

So a Congenital Defect is one that was present at birth. The name tells us nothing about the cause – whether it was genetic, environmental, traumatic, etc. The defect might not make itself known for days or years – but it certainly did not arise after the child was born.

NEW RESEARCH

Using this background knowledge of maternal risk factors, Dr Patrick Y. Jay and colleagues from the Washington University School of Medicine began their research.

They started with mice that had been specifically bred to have only one copy of the *Nkx2–5* gene (remember, it regulates the development of the heart). Because these newborn mice pups had only one copy of the gene (not the regular two copies), they had a 10 per cent risk of having a Ventricular Septal Defect, or VSD.

Specifically, the researchers looked at maternal age. With these specially bred mice, the risk of having a baby with a VSD increases further as the mother mouse gets older – but why?

To try to find out, this team looked at three possible influences that change with age – the age of the ovaries, the diet of the mother mice, and the exercise taken by the mother mice.

THREE POSSIBLE INFLUENCES

First, the researchers considered the age of the ovaries – and the eggs they carry.

Half of the female mice were young (equivalent to human teenagers), while the other half were approaching mouse menopause. The researchers transplanted young ovaries into the older mice, and old ovaries into the younger mice. The age of the ovaries had no effect on the number of VSDs in the baby mice. The young mice mothers (with the old ovaries) still had a lower risk of VSD. The older mice mothers (with the young ovaries) still had a higher risk. So the age of the ovaries and the eggs was irrelevant.

The second factor was the diet of the mother mice. As animals get older, their metabolic rate drops, fats and sugars are metabolised differently, and the waistline expands. The researchers fed the older pregnant mice either high-fat or normal kibble (mouse feed). Diet made no difference.

The third and final factor was exercise. Half of the young and old mice were given Running Wheels in their cages. They were allowed to run whenever they wanted for several weeks before becoming pregnant. The other half of the young and old mice had no Running Wheels – so they were the controls. Bingo!

Exercise was the key to fewer babies with Holes In The Heart. For the younger mice mothers, the risk was about 10 per cent whether they ran or not. For the older running mice mothers, the risk was kept down to about 10 per cent – but only if they exercised.

If they did not exercise, the risk of a Hole In The Heart jumped to about 20 per cent in the older mice mothers – they had double the number of pups with heart defects. (To make this kind of exercise routine relevant to humans, a dash on the Running Wheel for mice is equivalent to a brisk walk for you and me.)

STILL DUNNO

We still don't know exactly what it was about the Running Wheel that reduced the older mice's risk of having babies with a VSD.

We do know that exercise can change the function of some genes, and can also stimulate the release of many different chemicals from many different types of cells. Did some of these chemicals migrate into the bloodstream of the mother, and then into the growing mouse baby? Were there changes to the uterus as a result of exercise? We don't know – yet. (By the way, another limitation was this study didn't look at any effects of increasing age in the father.)

Furthermore, we don't know if this effect (the mother being able to "run down" the risk of a heart defect in her baby) works in humans – because last time I looked, mice are definitely not humans.

But certainly, when healthy pregnant women, and those contemplating pregnancy, become physically active there are many other potential health benefits for both mother and developing child. And there are very few disadvantages to exercise – as long as you don't overdo it.

So running with a baby bump,
might make a healthy pump,
for the baby you lump.

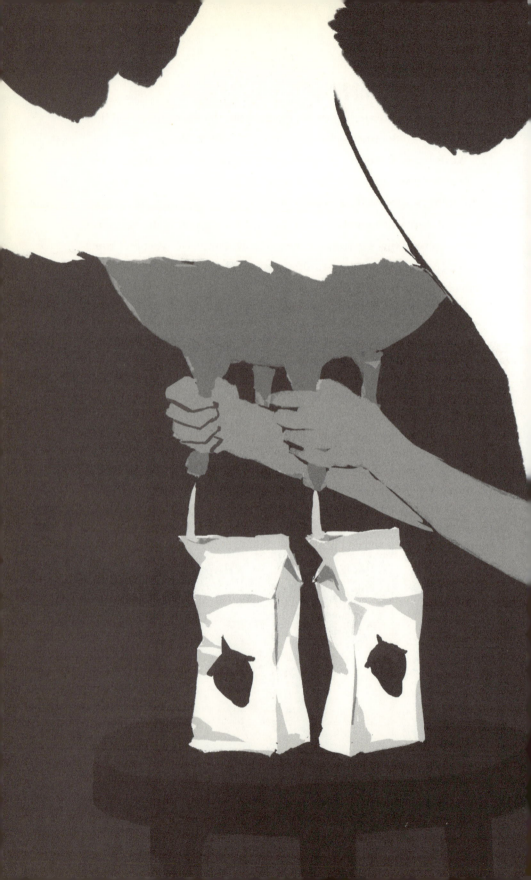

25

COWS MAKE FLAVOURED MILK

MY SECOND-FAVOURITE QUESTION FROM A CHILD (AFTER "WHY IS THE SKY BLUE?") IS, "IF COWS EAT LOTS OF STRAWBERRIES, WILL THAT MAKE THEIR MILK STRAWBERRY-FLAVOURED?"

The answer is, "The flavour of both cow and human milk can be changed by diet. Banana milk? Yes, we proved it. Strawberry milk? Dunno, yet..."

COWS' MILK

In France, the flavour of Gruyère cheese varies between summer and winter. It also changes if the cow has fed on the mountains, the plateaus or the plains. In other words, as the feed for the cow varies, so does the milk's flavour.

Back in Australia, on the Darling Downs, turnip weed can give the milk a rather "off" flavour. Farmers say they can taste a rather sharp radish-like flavour in the milk if their cows eat any of the cruciferous vegetables (e.g. cauliflower, cabbage, broccoli and Brussels sprouts).

In 1990, Australia's CSIRO wanted to understand how different feeds changed the flavour of cows' milk. Its researchers found that cows fed a diet low in fats produced what's called a "hard milk", with hints of blue cheese, coconut and peaches. In contrast, a synthetic diet free of protein gave a milk that was very low in stinky chemicals such as indole and skatole.

Then the CSIRO scientists found that if you supplement the feed of cows with oat and sunflower seeds simultaneously, you get an interesting surprise – via a two-step pathway.

First, the oats stimulate the growth of certain bacteria in one of the cows' stomachs. Second, these bacteria move on to the sunflower seeds. They convert the oil in the sunflower seeds into a chemical called gamma-dodec-cis-6-enolactone – which tastes just like raspberries. In fact, this chemical is so pervasive that it gives a raspberry flavour to both the cows' milk and their meat!

HUMAN MILK

The flavour of human breast milk is also affected by diet. Studies have found that if women drink carrot juice while they are pregnant or breast-feeding, their growing infants become very fond of carrot-flavoured cereal.

In 2008, Danish investigators at the University of Copenhagen fed some 18 nursing mothers four different types of flavour capsules. (Yes, I know, 18 isn't a very big sample size.)

Very quickly, the babies were enjoying flavoured milkshakes from their mother's breasts. It took varying times for the flavours to appear in the breast milk – one hour for banana flavour, two hours for caraway and licorice flavour, while menthol took between two to eight hours. After eight hours, most of the four flavours had disappeared from the mothers' breast milk.

EFFECTS OF FLAVOURED MILK

Flavours in breast milk can affect what a baby does. For example, we know that if the mother ingests garlic, alcohol or vanilla flavour, the baby attaches to the breast for longer.

Our adult food choices and dietary habits are partly genetic and partly environmental. We have learned many of our dietary habits from various cultural, psychological and physiological factors.

We already know that breast-feeding is linked with many good outcomes. Some are related to the immune system chemicals that the mother gives to her baby, others to the extended skin-on-skin contact.

But now we know that breast-feeding can also help baby develop a more sophisticated palate. It seems that the breast-fed babies quickly get used to small changes in flavour. This means they can become more accepting of a wide variety of flavours when they start to eat solid food. So changes in the taste of breast milk could prime the child to try different foods as they grow up.

What about non-breast-fed babies? They might get similar priming of their palates if they get fed a variety of different formula brands.

BACK TO STRAWBERRY MILK

So will feeding strawberries to cows produce strawberry-flavoured milk?

Dunno. I couldn't find a single case in the literature where a farmer had enough spare strawberries to feed them to cows.

But if the flavour does come through, perhaps strawberry milk could be the next totally organic, fully natural and permeate-free Superfood. And if chocolate worked, that would be Two-For-One...

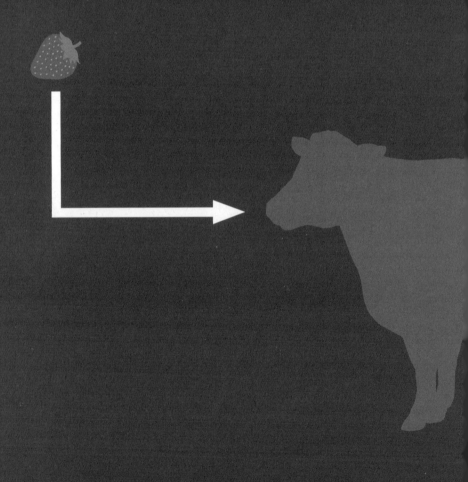

CHEMICALS OF FLAVOUR

There are two main ways that flavours or tastes get into milk.

In one case, the flavour chemical gets into the feed, and passes, unchanged, into the milk. This happens with garlic, and lucerne.

In the other case, chemicals in the feed (e.g. beet, rye and wheat) get changed by internal metabolism into other chemicals. It's these secondary byproducts that then make their way into the milk – and flavour it.

Beet byproducts contain a chemical called betaine. In the gut of the cow, betaine is converted into another chemical called trimethylamine, which gives a fishy flavour to the milk. If you feed rye or wheat to cows, different chemicals get converted to trimethylamine, and again the milk has a fishy flavour.

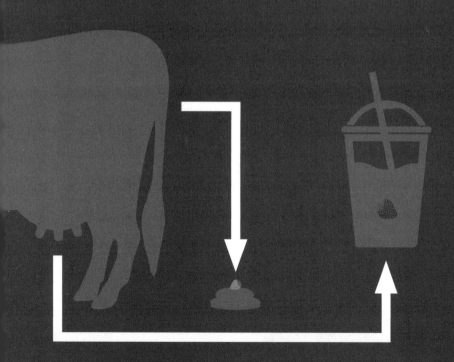

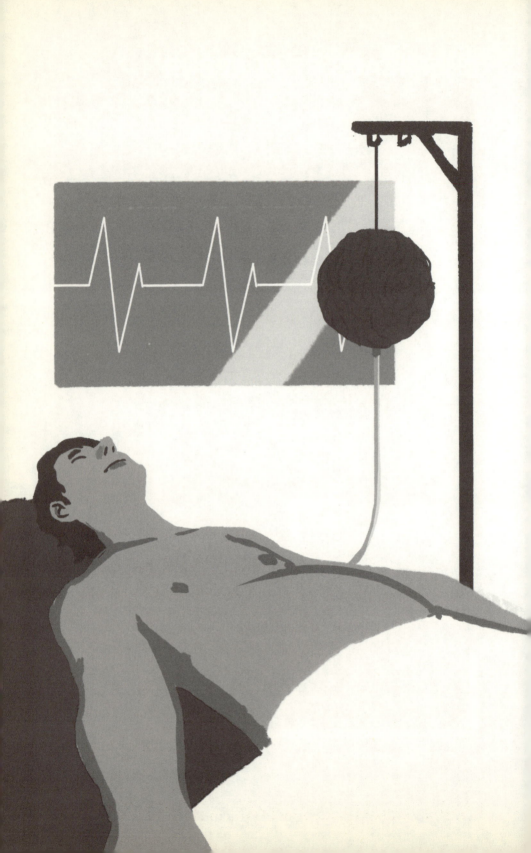

26

COCONUT WATER VS BLOOD

OVER THE PAST FEW YEARS, THANKS TO CLEVER MARKETING, COCONUT HAS BEEN "BLESSED" BY THOSE NOT TRAINED IN DIETETICS/NUTRITION WITH ALL KINDS OF OUTRAGEOUS AND UNPROVEN HEALTH CLAIMS. ONE OF THE MORE FRINGE CLAIMS IS THAT COCONUT WATER IS IDENTICAL TO HUMAN PLASMA. SO, BESIDES JUST DRINKING IT, YOU CAN INJECT THIS "PURE" HEALING NECTAR DIRECTLY INTO SUPPOSEDLY THE BLOODSTREAM.

Is there nothing that coconut cannot fix? Surely we should rush out and buy all products that are in any way "coconutty"? What about vampires? According to popular TV shows, they drink blood, so if coconut water "mimics" blood it must be good for both them and us?

(By the way, I wrote "Vampires Suck," a story on "real" vampires in my second book, *Even Greater Moments in Science*, way back in 1985. There's a medical link with the British Royal Family.)

Like all good myths, there's an element of truth buried inside the mistruths.

COCONUT WATER VS COCONUT MILK

Coconut water is very different from coconut milk. Coconut milk is an emulsion of freshly grated coconut flesh combined with coconut water.

COCONUT WATER 101

Coconut water is the liquid inside a young coconut – about 500 to 1000 millilitres per coconut. In the Solomon Islands, the coconut is an essential part of the diet. The locals describe six distinct stages of the coconut as it develops.

But to make it easy, let's assume a "young" coconut is about seven months old. That's when they carry the maximum amount of water relative to the coconut flesh, and also when the sugar level reaches its maximum.

If the shell of the coconut has not been cracked, the coconut water inside is usually "sterile" – that is, free of bacteria and the like. So if its constituents are very similar to those in blood, could it be injected safely into people, to replace fluid loss?

The answer is, "Maybe – if it's an emergency, and if the patient will otherwise definitely die. Otherwise, no."

ANECDOTES VS DATA

The plural of "anecdote" is "anecdotes" – not "data".

No matter how many times the same anecdote or story is retold, it does not turn magically into "data". It's just the same old story.

The tricky part is that we humans are "programmed" to readily accept statements from those we trust – even if they are merely inadvertently repeating mistruths they heard, in good faith, from other friends.

Richard Feynman (a Nobel Laureate and one of my heroes) put it nicely: "Don't trust authority, trust nature." In other words, don't trust stories – instead, trust what can be proven.

INJECTING COCONUT WATER

Back in 1942, Drs Pradera, Fernandez and Calderin in Havana, Cuba filtered coconut water. They then injected it into the veins of 12 children, at rates of around one to two litres per 24 hours. They reported no major adverse reactions.

It is also claimed that, during the Second World War, both the British in Sri Lanka and the Japanese in Sumatra regularly used coconut water when the standard intravenous fluids ran out. However, these claims are just anecdotes – they were never reported formally in the peer-reviewed medical literature.

In 1954, a group of doctors combined their findings. Between them, Drs Eisman, Lozano and Hager had administered coconut water intravenously to 157 patients in Thailand, the USA and Honduras. The vast majority of the patients, 136, were in Honduras. The doctors were dealing specifically with situations where genuine sterile medical infusion liquids could not be obtained.

Of 157 patients, 11 (about seven per cent) had reactions to the coconut water. These included fever, itchiness, headache and tingling in the hands. Furthermore, an "unspecified" number of patients also suffered aching sensations along the veins into which the coconut water was infused. This is thought to be due to the high potassium levels of coconut water.

COCONUT WATER IS NOT IDENTICAL TO BLOOD

	SODIUM (MEQ/L)	POTASSIUM (MEQ/L)
PLASMA	140	4.5
COCONUT WATER	3–9	35–80

And this brings us to the claim that coconut water is identical to blood. It isn't. ("They" lied.) In fact, in terms of the two major electrolytes (sodium and potassium), the levels are pretty well as wrong as possible.

Human blood is a suspension of water and cells. It's about 55 per cent salty water, and about 45 per cent cells. The cells are overwhelmingly red blood cells, with a tiny sprinkling of white blood cells and the like. The red blood cells give blood its red colour.

The salty water, called plasma, is a clear, slightly yellowish liquid, with high levels of sodium, low levels of potassium, and trace amounts of other minerals. As a result, genuine medical intravenous fluids are manufactured to have high sodium and low potassium.

Coconut water is the exact opposite – low sodium and high potassium. Coconut water has about one-twentieth the sodium level of plasma, while the potassium level is about 13 times higher.

But besides the high potassium, coconut water is also loaded with calcium and magnesium, which means it's definitely not suitable for patients with kidney failure, severe burns, etc.

Another problem is that it is much more acidic than human plasma.

The bottom line is that coconut water is not identical to human plasma – and it carries electrolytes in ratios that can be harmful.

MEDICAL USES OF COCONUT WATER

However, in an emergency (and let me stress *emergency*), coconut water can be used.

One case in 2000 involved a man in his forties who had recently suffered a stroke in the remote Solomon Islands. He had so much difficulty in swallowing that he choked on both liquids and solids and repeatedly vomited them up. He was rapidly dehydrating.

He was rehydrated initially in Atoifi Hospital with regular medical IV fluids. After a few days, a feeding tube was inserted into his stomach. This was now how he got his nutrition and fluids. After 36 days, he suddenly became weak, shaky and dizzy, and could no longer tolerate the feeding tube – so it was removed. Unfortunately, the hospital had run out of IV fluids, and because it was so remote, would not get supplies for another two days.

Over these next two days, because this was now an *emergency situation*, the doctors infused about two and a half litres of coconut water, to tide their patient over during the crisis. Luckily, he recovered the ability to swallow, and was discharged from hospital on Day 39.

So while coconut water is flavour of the month in Food Fad Land, taking it intravenously is one step too far – unless you're at death's door!

FICTION ...

I did a quick trawl on the web and found many, many mistruths about coconut water. Here are a few.

"COCONUT WATER IS IDENTICAL TO HUMAN BLOOD PLASMA WHICH MAKES IT THE UNIVERSAL DONOR."

"BY DRINKING COCONUTS, WE GIVE OURSELVES A [SIC] INSTANT BLOOD TRANSFUSION. IN FACT, DURING THE PACIFIC WAR OF 1941–45, BOTH SIDES IN THE CONFLICT REGULARLY USED COCONUT WATER – SIPHONED DIRECTLY FROM THE NUT – TO GIVE EMERGENCY PLASMA TRANSFUSIONS TO WOUNDED SOLDIERS."

"A PATIENT CAN SAFELY RECEIVE AS MUCH AS ONE QUARTER TO ONE THIRD OF THE PATIENT'S WEIGHT IN COCONUT WATER INTRAVENOUSLY."

VS FACT

NO, IT'S NOT. IT HAS THE WRONG RATIOS OF SODIUM AND POTASSIUM.

"REGULARLY"? NO. MAYBE A FEW TIMES – TOPS. DEFINITELY FEWER THAN TEN.

THERE IS NO LITERATURE TO SUPPORT THIS. UP TO 33 PER CENT! THIS CLAIM IS PLAIN CRAZY. LET ME PICK AN EXTREME EXAMPLE. IF A PATIENT IS IN SEVERE SHOCK (INVOLVING VERY LOW LEVELS OF BODILY FLUID), THEY MIGHT BE ABLE TO SURVIVE REPLACEMENT OF 10–15 PER CENT OF THEIR BODY WEIGHT AS MEDICAL IV FLUIDS. ANY MORE AND THEIR LUNGS WOULD BEGIN TO FILL UP WITH LIQUID, THEY WOULD "DROWN", AND MANY OTHER ORGANS WOULD GO INTO FAILURE.

27 WEIGHT OF A CLOUD

HOW MUCH DOES A CLOUD WEIGH?

It must have weight, because water has weight. A cloud is many, many tiny "clumps" of water, either liquid or frozen. The liquid droplets are about 0.002 millimetres across (smaller than the thickness of a human hair, which is about 0.050–0.070 millimetres). Some of these tiny droplets are so small that it would take a billion of them to make a single raindrop.

Different clouds carry different amounts of water. After all, cloud shapes and sizes can range from thin, wispy cirrus, right up to monstrous cumulonimbus thunderclouds.

A typical cumulus cloud carries about 0.50 grams of water (the weight of a big garden pea) in each cubic metre. The whole cloud might be one kilometre by one kilometre by one kilometre. So it could carry up to about 500 tonnes of water.

What holds this water up?

There are three main causes. First, the heat of the Sun warms the ground, which then creates rising currents of air. Second, an advancing storm or weather front running close to the ground can push the air that it runs into upwards. Third, air carrying water vapour can run into a mountain. The air rises and cools, which condenses the water vapour into water droplets.

Air (when it is moving very fast, relative to a small pair of wings) can easily hold up a 600-tonne jet plane. So a huge volume of slowly moving air can definitely hold up a 500-tonne cloud.

CLOUD TRIVIA

- The name "cloud" comes from the 13th Century Old English word "clud" or "clod", meaning a mass of rock, or a hill. A very dark cumulus cloud can look like a hill.

- Other planets have clouds. On Venus they are made of sulfur dioxide, while the clouds on Mars carry water as ice. The clouds on Jupiter and Saturn are (going from the top down) made of ammonia, ammonium hydrosulfide and near the bottom, water. The clouds on Uranus, Neptune and Titan (a moon of Saturn) are made of methane.

- The first successful weather satellite was TIROS 1, launched on 1 April 1960. It survived only 78 days, but proved that weather satellites could be useful.

28

VEINS ARE NOT BLUE

BACK IN YE BAD OLDE DAYS, PEOPLE OF NOBILITY WERE CALLED "BLUE BLOODS". YOU SEE, THE NOBILITY LED LIVES OF LEISURE, AND CERTAINLY DID NOT WORK IN THE FIELDS. AS A RESULT, YOU COULD EASILY SEE THE BLUE VEINS JUST UNDER THE SURFACE OF THEIR PERFECT, WELL-PRESERVED AND UNTANNED SKIN. ON THE OTHER HAND, THE PEASANTS HAD DARKER TANNED SKIN – WHICH MADE IT HARDER TO SEE THE BLUE BLOOD IN THEIR VEINS.

But why do the nobility, the peasantry, or indeed any of us, have veins that appear blue? After all, when you accidentally cut yourself, you see that human blood is red, not blue. (Blood from veins dribbles out, but blood from an artery can easily squirt half a metre or more.)

CARDIOVASCULAR 101

The heart is not the Organ of Love – it's a pump. In fact, it's a four-stage (or four-chamber) pump. Each stage (or chamber) accepts blood at a certain pressure, and then pumps it up to a higher pressure. It then gets pushed into the next chamber, or the lungs, or your general circulation.

Your blood (all five or so litres of it) is pumped around your body by this wonderful pump. (See the illustration in "Hole Hearted" on page 174). With a little care and maintenance, your heart should last over 70 years. In that time, it will have pumped some 170,000 tonnes of blood – roughly the weight of a giant ocean liner. Not bad for something weighing about 300 grams.

Let's start with the blood as it leaves the left ventricle of your heart at high pressure (say, 120 millimetres Hg) via the aorta. It's a light cherry-red colour, because it's loaded with oxygen. It travels via your arteries to pretty well everywhere in your body.

Oxygen is extracted and used by body tissues such as the brain, muscles, and even your skin. In the body tissues, the blood dumps its load of oxygen, while simultaneously picking up carbon dioxide.

As blood loses its oxygen, it changes in colour slightly, from light cherry-red to dark red. (Yes, it does get a little darker in colour. But this is not the only reason, nor the main reason, why veins *appear* blue.)

The blood is now both low in pressure, and low in oxygen.

Blood then flows through the smaller veins into the bigger veins, and ends up back in the heart – inside the right atrium (at about five millimetres Hg of pressure). This is the first chamber of the heart.

Then it is pumped into the right ventricle – where the pressure increases to about 25 millimetres Hg. This is the second chamber or stage.

It then enters the lungs where it regains both its oxygen and its light cherry-red colour. As the blood forces its way through the dense network of blood vessels inside the lungs, it loses about half its pressure.

So it enters the left atrium at the fairly low pressure of about 12 millimetres Hg. (This is the third chamber). This is not enough pressure to circulate through the whole body.

So the left atrium then pushes the blood into the left ventricle. (This is the fourth chamber). Here the pressure is increased back up to about 120 millimetres Hg – and the Circulation of the Blood is complete.

But at no stage does the blood turn a blue colour. It's always a shade of red – not blue. So why do veins look blue?

VEINS AREN'T BLUE

The first thing to realise is that that a naked, solitary vein is not blue at all. We know this thanks to surgeons. Surgeons see veins all the time. They tell us that, without the overlying skin, a vein carrying blood is not blue.

Veins are blue only when you look at them through the skin.

Why?

The best answer comes from a paper in the field of Optical Physics, written back in 1996. It was called "Why Do Veins Appear Blue? A New Look at an Old Question", and was written by Alwin Kienle and colleagues.

The blue colour of veins is caused by four separate factors.

They are that skin absorbs red colours, that de-oxygenated blood is less red, that deeper veins appear more blue, and that contrast of surrounding skin also "adds" blue.

NOBILITY

The word "nobility" comes from the Latin *nobilitas*, meaning "well-known, famous, notable".

The nobility made up varying percentages of the population in Europe over the centuries. Around the mid- to late-1700s they comprised 15 per cent of the Polish–Lithuanian Commonwealth, ten per cent of Castile, seven to eight per cent of Spain (in 1768), five per cent in the Kingdom of Hungary, but only two to three per cent of Russia (in 1760).

But, averaged across all of 18th century Europe, the nobility made up about two per cent of the population.

FACTOR 1: SKIN ABSORBS RED

The first reason is that skin tends to absorb the colour red.

Light interacts with the skin at different wavelengths or colours. On the way in, and then out of the skin, a lot of stuff happens to light – it will penetrate the skin, it will be absorbed, and finally, it will be re-emitted. This process of light being absorbed and then re-emitted happens many millions of times in the blink of an eye, as the light goes into and then out of the skin.

The physicists worked out their theory first, and followed it up with experiments. They filled glass tubes with blood and immersed them at varying depths in milk, which has a similar ratio of fats, proteins and water to human skin. They found that the veins near the surface re-emitted tiny amounts of red light, but lots of blue light. This meant that the colour blue was more noticeable.

FACTOR 2: DE-OXYGENATED BLOOD IS LESS RED

The second factor was how much oxygen the blood was carrying.

Most of the oxygen in blood is carried by very large molecules called haemoglobin. Fully loaded, a single molecule of haemoglobin can

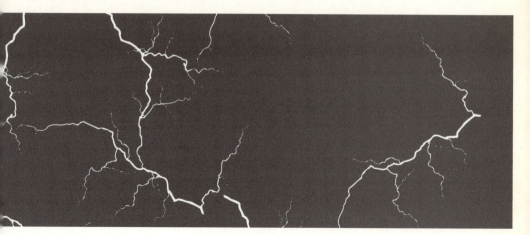

carry four atoms of oxygen. But if the conditions are right (e.g. high temperature, acid environment, and so on), one or more of the oxygen atoms will leave the haemoglobin.

As the oxygen level of the haemoglobin goes down (four oxygen atoms to three, then two, then one, then zero), the colour of blood changes from light cherry-red to a darker red.

Dark red is still red, but getting closer to our mysterious blue.

PART 3: VEINS, DIAMETER AND DEPTH

The third factor is how deep the veins are – the deeper they are, the bluer they appear.

If the vein is immediately under the skin, it will appear less blue.

The overwhelming majority of veins are deeper than 0.5 millimetres. Below 0.5 mm, thanks to the complicated Optical Physics involved in the so-called "Transport Equation", the vein will look more blue.

There's another factor. Remember that de-oxygenated haemoglobin in veins is a slightly darker shade of red than the oxygenated haemoglobin in arteries. This rather small difference is amplified as the light travels through the skin.

The overall result is that, comparing arteries and veins, the veins will look more blue. And because the arteries are mostly smaller in diameter and deeper down, they will usually not be seen at all.

With all this talk of blood, I hope you squeamish readers aren't turning a whiter shade of pale . . .

FACTOR 4: VISUAL ILLUSION, OR NEARBY COLOURS

Finally, there is the fourth factor – your brain.

Your brain does a huge amount of processing on the electrical signals that come in from the retina.

For example, how many grey dots can you see on the opposite page?

This is called the Hermann grid illusion where your brain perceives "ghostlike" grey dots of a white grid on a black background, but they disappear every time you look at them directly. In the case of veins under the skin, the contrast of the surrounding skin tends to make the veins appear more blue in colour (the physics is complicated and involves big equations).

For another weird and wonderful optical illusion, see the endpapers on the front and back of this book.

29

BACTERIO-PHAGE & L'ORÉAL

IN 1896, WHILE IN INDIA, THE ENGLISH CHEMIST ERNEST HANBURY HANKIN INVESTIGATED THE SUPPOSED HEALING POWERS OF THE INDIAN HOLY RIVERS. HE DISCOVERED SOMETHING THAT COULD KILL THE CHOLERA BACTERIUM. THIS MYSTERIOUS AGENT COULD BE DESTROYED BY HEAT – AND WAS FILTERABLE. (IT WAS ABLE TO PASS THROUGH A FINE PORCELAIN FILTER THAT WOULD STOP TINY BACTERIA.)

POCKET PENICILLIN

Howard Florey and his team worked on mass-producing penicillin during the Second World War. They knew the importance of their work. But they, and their laboratory, were under continuous risk of being bombed.

They were worried – what would happen if their laboratory was bombed into oblivion? And what

That meant that this "mysterious agent" was smaller than bacteria – it turned out to be about one hundredth the size. This was our first scientific inkling of the existence of the bacteriophage.

The inklings were two independent discoveries.

The British bacteriologist Frederick Twort, Superintendent of the Brown Institution in London, described in 1915 how he grew bacteria called *micrococci* on agar culture plates. But when he added a "mysterious agent" to the agar plates, the bacteria would die – they would undergo "glassy transformation". He suggested that this agent could have been giving the bacteria an "acute infectious disease", which killed them.

In 1917, French-Canadian microbiologist Félix d'Herelle, at the Pasteur Institute in Paris, observed the same phenomenon. He coined the word "bacteriophage". He thought of it as being a "virus parasitic on bacteria".

His next step was to commercialise preparations of bacteriophages as a curative therapy for various infectious diseases. They had names indicating where, and for what, they should be used – Bacté-coli-phage, Bacté-pyo-phage, Bacté-rhino-phage, Bacté-intesti-phage, and Bacté-staphy-phage. He teamed up with the Société Française de Teintures Inoffensives pour Cheveux (Safe Hair Dye Company of France) – now better known as L'Oréal.

So it was a hair company that first commercialised phages.

would happen if most of the team was also bombed into oblivion?

So, at the end of every day, each of the team members would get some of the penicillin culture – and smear it onto the inside of their trouser pocket. That way, the precious penicillin would survive.

PHAGE'S GLORY DAYS

By 1921, the first therapeutic trials of phages had begun in Paris. When phages were injected into the staphylococcal boils of six patients, there seemed to be improvement. There also seemed to be improvements in small-scale trials when Félix d'Herelle gave phage treatments to patients with cholera, bacillary dysentery, and bubonic plague. On a larger scale, he was credited with using phages to stop outbreaks of cholera in India, and of the plague in Egypt.

Félix d'Herelle's report of success with the hitherto-impregnable plague stirred the imagination of the author Sinclair Lewis. In 1925, Lewis wrote the book *Arrowsmith*, in which the hero, a medical doctor, used phages for the common good – but suffered personal disaster. In 1926, Lewis was awarded the Pulitzer Prize for his book, but he refused to accept it.

By the 1920s in the United States, Western pharmaceutical companies such as Eli Lily were using bacteriophages to treat bacterial infections. In 1923, in Tbilisi, Georgia (in the then-Soviet Union), George Eliava opened the modestly named Eliava Research Institute to investigate bacteriophages. At the time, there was nothing else that would safely work against bacterial infections – antibiotics had not yet been invented. Millions of people were treated with phage therapy.

GLORY DAYS FADE IN THE WEST

The mainstream acceptance for phage therapy began to fade in the mid-1930s.

The rise of antibiotics made phages look like a slow, cumbersome and labour-intensive solution to a problem already solved.

Sulfonamide, the first antibiotic, became available in the early 1930s. It worked only moderately well – so phages still had a role to fill.

But soon, an even better antibiotic appeared – penicillin, the Magic Bullet. Back in 1928, Alexander Fleming had returned from a two-week holiday to find the now-famous "clear spot" on his dish of bacteria. But he was notoriously poor as an author and communicator – and word of his discovery spread very slowly indeed. It took until around 1940 for an Australian, Howard Florey, to work out how to mass-produce penicillin.

In the West, bacteriophages were abandoned immediately after the widespread introduction of antibiotics in the early 1940s. There were several reasons for this.

First, antibiotics were cheap, as well as being easy to make, store and prescribe. In contrast, bacteriophages had to be individually blended, in response to a specific demand. Their effectiveness varied from person to person, depending on what particular sub-species of bacteria was attacking then. If you chose the wrong phages, they didn't work.

Second, antibiotics could attack a whole range of different bacteria. The treating doctor needed to prescribe only penicillin. Bacteriophages, however, would work only against very specific bacteria. If you wanted to kill a different bacterium, you needed to use a different bacteriophage, or mix of bacteriophages. So bacteriophages were both unpredictable and fastidious.

Third, back in the early days, there were major problems with the medical trials and evaluations of bacteriophages. Some of the studies were carried out on diseases for which bacteriophages definitely would not work, such as viral infections or allergies. (Bacteriophages attack only bacteria – and neither viruses nor immune system cells are susceptible, because they are not bacteria.) Furthermore, some of the studies had poor controls, or none at all. Even worse, some so-called "phage therapies" carried no phages at all.

To be fair, it was only in the 1940s (a decade after the invention of the Electron Microscope) that we proved positively that bacterio-phages really existed – and also proved they were a virus. If you were preparing a phage therapy solution, it must have been hard dealing with something one-hundredth the size of a bacterium – and thus totally invisible.

BAD SCIENCE

In 1932, one American health officer presciently warned, "Because of conflicting experimental observations, enthusiastic and poorly controlled clinical application and rapidly expanding commercial exploitation, a situation is developing which will, unless guided and checked, lead to the ultimate rejection of bacteriophage by all who make any pretense to the practice of scientific medicine."

Unfortunately, Bad Science did hinder the advance of Bacteriophage Therapy.

GLORY DAYS CONTINUE IN SOVIET UNION

There was a political factor involved in the continuing use of phages in the Soviet Union.

After the Second World War, the Iron Curtain descended and the Cold War began. Part of the economic sanctions against the Soviet Union by the West included an embargo on the importation of antibiotics. So in Georgia, the development of bacteriophages continued unabated – even until the present day. They had no alternative. Phage therapy remained popular in Georgia, Russia and Poland.

However, there was a fundamental problem – the mediocre quality of much of the Soviet research.

ELIZABETH TAYLOR AND BACTERIOPHAGE

In March 1961, Elizabeth Taylor was in London filming *Cleopatra*. She caught pneumonia, had an emergency tracheotomy – and then got worse.

A Philadelphia-based company, Delmont Laboratories, sent "staphyloccus bacteriophage lysate" to London at the specific request of Miss Taylor's doctors.

It wasn't until the mid-1970s that the Soviets began to conduct high-quality research in phages. Unfortunately, nobody translated the research from the Russian, Georgian or Polish languages into English. So even in the 1970s, very few people in the West had even heard of bacteriophages.

The Berlin Wall fell on 9 November 1989. As the Soviet Union collapsed, scientific communication crossed the old Iron Curtain. Finally, the good word on bacteriophages began to spread.

PHAGE PLANS

Bacteriophages have been accepted in the West very slowly.

It took until 2006 before the American Food and Drug Administration finally approved bacteriophage products for humans. (Mind you, we eat millions of them with each mouthful, breathe in millions more with each breath – and there are trillions inside us already.) Besides being used to treat ear infections in dogs, phages are also now being used as a food additive to be sprayed onto ready-to-eat meats, to kill both *E. coli* and *Listeria* bacteria. In the United States alone, *Listeria* causes about 2500 cases of severe food poisoning, and kills some 500 people each year. The product is called ListShield, and is a cocktail of six different phages.

Her unusual phage therapy may or may not have been instrumental in her recovery. (We have no way of telling – no controls, small sample size of one, etc.)

Today, Delmont Laboratories no longer caters for the human phage market. But it makes a USDA-licensed phage therapy for recalcitrant skin infections in dogs.

In 2011, the American Food and Drug Administration approved a bacteriophage-based diagnostic product. It can distinguish *Staphylococcus aureus* from other bacteria, and determine if it's the nasty methicillin-resistant variety known as MRSA. It has reduced time to determine Microbial Identification and Susceptibility from up to 72 hours down to just five hours. This time will be reduced further with future improvements.

In March 2014, the US National Institute of Allergy and Infectious Diseases recognised the importance of phages. They listed phage therapy as one of its pathways to combat antibiotic resistance in bacteria.

In May 2014, the European Union committed an initial €3.8 million to the Phagoburn research project. This project will recruit 220 burns victims whose burns have become infected with *E. coli* or *Pseudomonas aeruginosa*. The bacteria will be treated with over 1000 phages – from sources such as river water or sewage.

So here is a good-guy virus, unlike viruses such as SARS, MERS and Ebola. How shall we grow together over the next few decades?

30

EARWAX & ARMPIT SWEAT

HERE'S THE TAKE-HOME MESSAGE: WET EARWAX AND WET ARMPITS GO HAND-IN-HAND. SAME FOR DRY EARWAX AND DRY ARMPITS.

AN EARDRUM CONVEYOR BELT?

We discovered the "Eardrum Conveyor Belt" back in 1964 – by placing tiny spots of dye on the eardrum and tracking their motion.

The "skin" of the eardrum spreads outwards from the centre of the drum, at roughly the rate your fingernails grow. (Fingernails grow at an average speed of

---

Whether your earwax and armpits are wet or dry depends on your genetics. Some peoples (such as most Europeans and Africans) sweat a reasonable amount. They tend to have wet earwax. And some peoples (such as most Koreans and Chinese) don't sweat much at all. They tend to have dry earwax.

But here's something weird. About two per cent of Europeans have dry earwax and don't sweat much. It turns out that about three quarters of this group use underarm deodorants, even though they don't need them.

EARWAX 101

Earwax is manufactured inside your ear canal. It's made from dead skin cells, as well as the secretions from two types of glands – sebaceous glands and sweat glands.

These glands are mostly in the outer part of your ear canal. Earwax is made from triglycerides, free fatty acids, cholesterol, a whole bunch of other fats and waxes, amino acids, minerals, and dry flaky skin. Earwax both cleans and protects the ear canal.

But ear doctors have long known that there are two quite different types of earwax – wet and dry. Wet earwax is sticky, and light-to-golden brown in colour – although it can darken over time. Dry earwax is flaky and brittle, and ranges in colour between brownish grey and light grey.

about three millimetres per month.) As the eardrum "skin" gets further from the centre, it accelerates. The skin then migrates along the wall of the canal, heading outwards.

As the skin of the ear canal drifts outwards, it carries with it earwax, as well as foreign bodies such as dirt and dust.

GENETICS OF EARWAX

The genetics behind wet versus dry earwax was discovered in 2006. It follows remarkably simple Mendelian genetics (as taught in schools, e.g. Cystic Fibrosis).

Dry earwax is very common (80–95 per cent of the population) among East Asians, but is less frequent (30–50 per cent) in southern Asia, Pacific Islands, Central Asia and Asia Minor, and in Native North Americans and Inuit peoples of Asian ancestry. It's very uncommon among Europeans and Africans (0–3 per cent). They overwhelmingly have the wet earwax.

The type of earwax you have – wet or dry – is caused by a single variation in your DNA, at a specific location on your 16th chromosome. (It's called the ABCC11 genotype.) As a result, in your ear canal, your glandular cells either make wet earwax – or they don't. This variation in the DNA turns out to have effects on the functioning of your liver, pancreas, kidneys, placenta (in pregnant females only), breast tissue, gut, glands in your ear canal, and wait for it, sweat glands in your skin – and that includes your armpits.

And yes, modern Genetic Knowledge tells us that humans with wet earwax will sweat more from their armpits.

DEODORANTS

The annual US market for deodorants and antiperspirants is worth about US$2 billion, while the UK market runs to over half a billion pounds sterling. In a study of over 18,000 people in the UK, about 93 per cent of women and 83 per cent of men used an armpit deodorant nearly every day.

So the tiny minority (about two per cent) of Britons who have dry earwax and so do not sweat much from the armpits surely don't need to use a deodorant? And yet, according to a study in 2013, about three quarters of them do use a deodorant daily. This shows the power of social conditioning and relentless marketing.

Perhaps there's a whole new potential market out there promoting products to treat sweaty ears.

31

MICE ON WHEELS

YOU'VE PROBABLY HEARD OF THE "HAMSTER WHEEL" OR "RUNNING WHEEL". IT'S AN EXERCISE DEVICE USED EITHER FOR ANIMALS AT HOME OR IN THE LABORATORY – MAINLY HAMSTERS AND OTHER SMALL RODENTS.

But it turns out that creatures in the wild love to have a go on it as well.

RUN FOR SCIENCE

Animal scientists have long used these Running Wheels to work out the effects of exercise on general health, cancers, circadian rhythms, sleep disorders, diabetes, depression, growth of new nerves in the rodent brain, and so on. (See "Hole Hearted" on page 173.)

In the USA alone, about 15 to 20 million lab rats and mice are given the "opportunity" to "volunteer" for laboratory testing each year.

But there has always been a persistent underlying concern.

Were the animals running on the wheel because they were unhappy to be in a cage? Perhaps the running was actually a Neurotic Repetitive Behaviour, which was subtly alerting us to the fact that the animals were deeply disturbed?

After all, in the aftermath of a major disaster or personal calamity, you might see people rocking back and forth as their way of coping with the stress. It's called "Stereotypic Behaviour" – it's repetitive, unchanging and seems to have no obvious goal. On the other hand, it does seem to help people deal with stress.

LORENZ – BIRD IMPRINT MAN

Back in 1967, Konrad Lorenz passed on a snippet of knowledge to a researcher in the field of Animal Behaviour. Lorenz was an Austrian zoologist, ethologist and ornithologist who jointly won a Nobel Prize in 1973.

He is best popularly known as the scientist who discovered "imprinting". Some birds that leave their nest early (such as geese) will "bond" very strongly with the first moving object that they see within the first hours of hatching from the egg. If that object is you, then you are their "parent" forever after.

So what was the snippet of knowledge? Lorenz once told scientist J. Lee Kavanau that laboratory rodents who had used Running Wheels, and had later "busted out" to freedom, would re-enter and run in one of these wheels quite voluntarily if they (randomly) came across one.

RUN FOR FUN

Professor Johanna H. Meijer and Yuri Robbers from the Leiden University Medical Center in the Netherlands wanted to know if animals "enjoyed" Running Wheels.

So they took two Running Wheels out of the laboratory and into the Great Outdoors.

One set-up was in a "Green Urban Area" – which is fancy scientific talk for "Professor Meijer's backyard". The other was set up in a sandy dune area that the public did not have general access to.

Each of these Running Wheels was inside a cage with a small door, so only small animals could get in. The scientists put food inside the cage to attract the local wildlife. In each case, an infrared camera, activated by a motion sensor, was set up to record any Running Wheel activity. The animals set off more than 200,000 video recordings, but only about 12,000 of these deserved deeper analysis.

BORN TO RUN

Some relatively small animals (i.e. with short legs) can cover huge distances in the wild.

Rats have been clocked as covering 43 kilometres in 24 hours. In a single day, wild mice have run 31 kilometres, lemmings 19 kilometres, and laboratory mice 16 kilometres, with gerbils coming last at eight kilometres.

PARTY WITH FOOD

About 1000 animals voluntarily went for a run on the wheel in the backyard (over a period of over 40 months, with a peak in summer), and about 250 on the wheel in the dune area (over 19 months, with a peak in late autumn). They would mostly drop in at night.

There were more juveniles than adults running on the wheels. The creatures were overwhelmingly wild mice. But a few rats, shrews, frogs, birds, snails and slugs also dropped in to do some running, hopping and slithering. In fact, slugs were the second most frequent visitor. Sometimes the slugs would slither along for hours.

The birds didn't do any running – they just had a look. Frogs hopped up and down, making the wheel move back and forth.

The snails were too haphazard and disorganised to keep moving in the same direction. They also "cheated", according to Professor Meijer. "Snails in the wheel . . . activated the wheel without running. They climbed up vertically until the centre of mass was moved about the central axis of the wheel, making it turn."

PARTY WITHOUT FOOD

Then the scientists wondered whether maybe the animals had been attracted by the food. Perhaps the animals were under the delusion that if they kept on running, more food would be forthcoming.

So the scientists stopped providing food.

But over the next 16 months the animals kept on dropping in for a run. There weren't as many, but they still came. About half of the drop-ins were very young mice that had been born since the researchers stopped providing food. So they were definitely dropping in with the specific goal of running on the wheel. It was their free neighbourhood Mouse Gym – no joining fee, and well maintained.

Of course, not all wild animals did it deliberately. Some of them would get on, check out the Running Wheel, decide it was not for them, and never come back. But many were proper little Gym Junkies, and they knew what they were doing.

They'd run for a bit, get off for a rest, and then hop back onto the Running Wheel again. In fact, the wild mice were taking their local gym quite seriously. On one hand, their average speed of running was a bit slower than that of the lab mice – 1.3 kilometres per hour as opposed to 2.3 kilometres per hour. But on the other hand their top speed of 5.7 kilometres per hour was higher than lab mice ever reached – a mere 5.1

kilometres per hour. Apparently, the wild mice liked to practise their sprinting.

"Mice run in the wheel; they never just walk. They are frequently observed leaving the wheel, and immediately going back in, suggesting it is a voluntary act," said Professor Meijer. "Sometimes two mice run in the wheel at the same time."

IS IT EXERCISE IF IT'S FUN?

Now, we know that about two thirds of early deaths in humans are related to lifestyle – too much food, not enough exercise, and the like.

We also know that exercise helps you grow new brain cells, slows ageing, and reduces the incidence of various cancers, diabetes, heart disease and stroke. In general, exercise is good for you. (Of course, exercise should be planned to avoid injuries – both short-term and long-term.)

At this stage, all we know for sure is that wild mice will go for a run on the exercise wheel – but we don't know why.

Did the animals "know" that exercise is good for them? Were they getting the so-called "Runner's High" from endorphins or endocannabinoids? Were they working on their thighs? Or were they simply playing, and having fun?

In the case of playgrounds for human children, there are plans to include equipment modelled on exercise wheels. Adults are most welcome to hop on for a few spins. Do we hop on for exercise, or for play, or because we enjoy unexpected movement?

I wonder if the scientists would get more mice on their Running Wheel if they installed mirrors in the cage, and began blasting "doof-doof" music?

32

TOILET SEAT GAVE SOUND RANGING

IN THE FIRST WORLD WAR, THE SINGLE MOST EFFECTIVE WEAPONS SYSTEM (OR KILLING MACHINES, TO PUT IT BLUNTLY) WAS ARTILLERY – THE BIG GUNS. THE BEST WAY TO DEFEND YOURSELF FROM ENEMY FIRE WAS TO DESTROY THE ENEMY'S ARTILLERY.

But first you needed the enemy guns' exact location. The most successful technology at pinpointing enemy artillery was "Sound Ranging". A team led by a 25-year-old Australian Nobel Prize winner and serving soldier, William Lawrence Bragg, developed this amazing technology during the war.

And what was Lawrence Bragg's "Ah-ha" moment of inspiration? He realised, while sitting on the toilet, that his bare bottom was getting lifted off his toilet seat whenever a nearby artillery piece was firing – even when he couldn't hear the gun being fired.

ARTILLERY 101

The First World War ran from mid-1914 to late 1918 – about four and a half years. Approximately nine million military personnel and seven million civilians died in that bloody conflict. On top of that, over 22 million military personnel were seriously injured or disabled.

Overall, during the course of the First World War, exploding shells and shrapnel fired by artillery killed more soldiers than any other weapons system.

Artillery at this time was very inefficient – more than 100 shells had to be fired to kill just one soldier. But it was terribly effective, killing millions of people during the course of the war. (Hundreds of millions of artillery shells were made and sold to the armies of both sides – and huge profits were made by weapons manufacturers.)

FINDING THE BIG GUNS

At the beginning of the First World War, neither side had effective and safe methods that could pinpoint the location of enemy artillery.

Sure, they had flash-spotters. These were brave men who put themselves in locations where they could see the flash of light from the tip of the artillery barrel (the "muzzle flash") as the big gun fired. But if the spotters could see the artillery piece, then the operators of the big gun could potentially see them. Often the big guns were hidden out of sight, which made it very difficult to spot the flash.

Another method was to use the recently developed aeroplane for spotting big guns. But planes couldn't do their spotting in bad weather, or at night.

MEASURE ARTILLERY VELOCITY

The military always want their artillery shells to land where they aim them. They need to know the "muzzle velocity" – the speed at which the artillery shell leaves the barrel. But it's hard to measure.

William Lawrence Bragg hung two webs of wire in the flight path of the artillery shell. He then ran electricity into each web of wire. As the artillery shell ripped through first one, then the other web of wire, it generated electrical signals that Lawrence Bragg measured, giving him the muzzle velocity.

And how did he record this? Simple – he just painted the muzzle velocity onto the gun barrel.

GEOMETRY & SIGHT

So come on down, Sound Ranging.

This method was based on an ancient geometry technique called Triangulation (a fancy word meaning "geometry with triangles").

The history of Triangulation goes way back. About two and a half thousand years ago, the Greek mathematician and philosopher Thales was using a technique called Similar Triangles to estimate the height of the Pyramids of Egypt. Around 250 AD, the Chinese scientist Pei Xui was using triangles to make accurate maps.

Surprisingly, the Romans don't seem to have used Triangulation. However, around 1000 AD, various Arabian scholars were so familiar with Triangulation that they were using it to measure distances, make maps and even measure the size of the Earth. It took a few centuries for Arabian Mathematics to slowly percolate across to the West. Eventually, in 1533 the Dutch cartographer Gemma Frisius wrote about using Triangulation to accurately work out the exact position of distant locations.

The principle is simple. Suppose there is a mountain peak, clearly visible in the distance. You mark out a baseline, say, two kilometres long. You walk to one end of your baseline, and measure the angle to your mountain peak. You then walk to the other end of your baseline and measure the new angle to your mountain peak. Then, with these two angles and the baseline, simple geometry can provide the exact distance to your mountain peak.

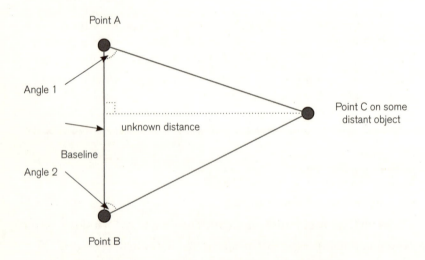

But everything is messy on the battlefield, in the fog of war. First, it's quite difficult and dangerous to walk around with a set square and protractors when people are shooting at you. But the major problem is that you can't reliably see where the enemy artillery is.

However, you can usually hear it.

GEOMETRY & SOUND

We know that sound covers about 300 metres each second. On one hand, you can easily pick up the sound of a big gun firing with a microphone. However, on the battlefield, you usually cannot see the muzzle flash at the exact moment it fired. This means that you don't know when to start your timer. As a result, you don't know how long it is between the gun firing and the sound arriving at your microphone.

Luckily, there is a way around this.

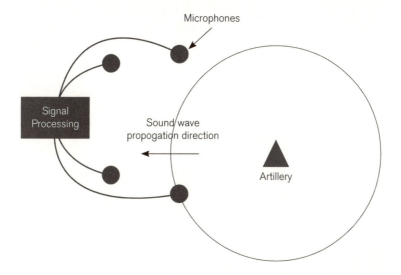

Suppose you have a string of microphones located four kilometres back from the front line, and spread along a baseline nine kilometres long. The sound from the big artillery piece firing off arrives at different microphones at different times. If you have enough microphones along a long enough baseline, you don't need to see the muzzle flash. If you can record the timing of when the sound arrives at each microphone, basic maths gives you the exact location of the enemy artillery – regardless of the weather and the time of day.

That all sounds fine – except for another fundamental problem. Sometimes, soldiers can't hear the enemy artillery piece firing – because of its location, and the general din of war.

BRAGGS' NOBEL PRIZE

In 1912, Lawrence Bragg was walking along the banks of the river Cam (hence the name "Cambridge"). That day, Lawrence had a sudden insight into the nature of X-Rays that he took to his father, William Henry Bragg, the Professor of Physics at Leeds University.

Father and son worked together for two years. Lawrence remembered their collaboration as "a glorious time when we worked far into every night with new worlds unfolding before us in the silent laboratory."

In 1914, he and his father jointly published their book, *X-Rays and Crystal Structure*. In May 1915, they were jointly bestowed the Barnard Medal of Columbia University. This prize was awarded at five-year intervals, as an "American Medal for British Savants". Mind you, William Lawrence Bragg was born and educated in Australia, and Australia had been granted its independence in 1901 – so we can argue that he was True Blue Aussie.

Their Nobel Prize was awarded on 12 November 1915. In their Nobel Lecture, belatedly given in 1922, Lawrence Bragg suggested, "There seems to be hardly any type of matter in the condition of a true solid which we cannot attempt to analyse by means of X-Rays".

By that time, the science of Chemistry had given us methods to work out what kinds of different atoms were in a chemical (e.g. carbon, hydrogen, oxygen), and how many of each of them were present. But Bragg's work took us one step further. It let us work out how these atoms were arranged. So you could blast a beam of X-Rays into a crystal, measure how much the beam was bent, and from that information work out the locations of the atoms inside that crystal. Understanding the structure of DNA, one of the most profound discoveries about the nature of Life we humans have ever made, was obtained using the knowledge they had uncovered.

228 > Dr Karl's Short Back & Science

And that's the problem Bragg solved. The 25-year-old Australian Nobel Prize winner, serving at the front, got lifted off the toilet seat by the firing of the big guns just once too often...

NOBEL PRIZE WINNER ON THE BATTLEFIELD

In September 1915, during the second year of the First World War, William Lawrence Bragg was serving in France. He had just heard the terrible news that his brother, Bob, had died fighting at Gallipoli. Soon after, he received a much happier letter from his father telling him that they had been jointly awarded the Nobel Prize in Physics.

They are the only father–son team to share a Nobel Prize. William Lawrence Bragg is still the youngest Nobel Laureate in Physics – he was just 25 years old.

He had gone to Trinity College at Cambridge in 1909 to read mathematics. The same year, he enlisted in a Territorial Force formation, King Edward's Horse. This had been set up specifically to train men from the British Colonies. Certainly, it was a great way to meet people like himself – strangers from afar, and slightly out of place attending a very old university. He was discharged in November 1913. The First World War broke out in August 1914 – and by 26 August he had been commissioned as a Second Lieutenant in the Leicestershire Royal Horse Artillery. He started the war riding horses in the Mounted Infantry.

In July 1915, with absolutely no warning, Bragg was ordered to report to the Maps division of General Headquarters. The Allies were trying to pinpoint the position of German artillery pieces using various technologies. The most promising technique involved recording the boom of artillery with an array of microphones – simple in concept, but difficult to implement.

NOBEL PROFESSION

Lawrence Bragg loved gardening. He missed it awfully when he moved to London. So he worked as a part-time gardener, totally unrecognised by his new employer.

However, he got sprung when a guest of his employer recognised him, and expressed surprise at seeing a Nobel Laureate doing the gardening.

HEARING ARTILLERY

The problem was that they didn't always hear the big guns.

When a cannon fires, it generates mostly low-frequency sounds. A smaller field gun booms at 25 hertz, but a larger artillery piece emits most of its sound energy as inaudible infrasound – way down at 10 hertz. Back then, nobody knew this. Furthermore, the microphones of the day would barely respond to such low frequencies.

They needed to know that the low-frequency infrasound existed – and how to capture that energy. The breakthrough came as two separate hints from Nature.

Hint Number One was very much related to a Call of Nature! Bragg was billeted in accommodation in Flanders. The toilet was a small room, with a door, but no window. When the door was shut, the only connection to the outside world was the pipe leading from under his toilet seat.

By fortunate happenstance, there was a British 6-inch artillery piece about 400 metres away. It was just at the "right" distance. When it fired, sometimes Bragg could hear it, and sometimes he couldn't. But each time it fired, his bare bottom was actually lifted off the toilet seat by the inaudible infrasound energy – even though sometimes he would hear nothing at all. The infrasound energy of the firing was being channelled through the pipe leading to his toilet bowl.

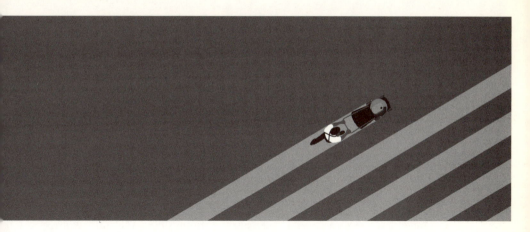

So now he knew there was enormous energy in the inaudible infrasound – but he didn't how to detect it.

Hint Number Two came via Corporal W.S. Tucker, also a Physics graduate, who served in Bragg's team. Tucker was not an officer, and so his accommodation was less salubrious. He slept in a shack that had walls, not of wood or brick, but tar-paper – literally, stiff paper that had been covered with tar to make it waterproof. There was a hole in the tar-paper next to his pillow.

When he heard the 6-inch gun fire, an annoying puff of chilly air would strike his cheek. But every now and then, even though there was no wind blowing, and he had not heard that 6-inch gun fire, he would suddenly feel that puff of air.

DETECT THE INFRASOUND

The two physicists realised that when a big gun fired, its high-energy, low-frequency infrasound could push air with quite a lot of energy. This moving air could lift a bare bottom off a toilet seat, or push through a hole in a shack's wall.

To the physicists, the solution was obvious. All they had to do was reliably detect a puff of air.

They got a small, empty wooden ammunition box, and ran a thin platinum wire from one end to the other, inside the box. They then drilled a hole in the box right next to the wire. They ran electricity through the wire, which heated it up. When the 6-inch gun fired, the infrasound pressure wave forced a puff of air onto the hot wire and cooled it down. This changed the resistance of the wire – something that was very easy to measure with the technology of 1915. They had just invented an Infrasound Microphone. They recorded this signal from their Infrasound Microphone onto smoked paper.

Once Bragg's team proved that their technology could actually work, their next job was to make it work reliably under the harsh conditions on the battlefield. After much development, Allied Sound Ranging could locate German artillery to within 25 to 100 metres – depending on various factors.

The Germans never worked out this technology during the course of the war. They did, however, know that the English had a superb technology for locating their guns. A captured German order read, "In consequence of the excellent Sound-Ranging of the English, I forbid any battery to fire when the whole sector is quiet . . ."

Sound-ranging was brilliantly demonstrated on 20 November 1917, in an assault along the Hindenburg Line near the German-occupied French city of Cambrai. This Battle of Cambrai was the first successful use of a large number of tanks in battle. Thanks to the new technology of Sound Ranging, the German artillery was quickly silenced. It was a great hour for the Allies – church bells were rung in London. Sound Ranging again was used to devastating effect in the Battle of Amiens, on 8 August 1918. And shortly thereafter, the war was over.

I guess that proves, once and for all, that men do their best thinking on the toilet seat, and should be allowed to sit there for as long as they like . . .

WHICH SON?

The father, Henry Bragg, was rather careless in not actually naming his son and collaborator, Willian Lawrence Bragg, in their research. Henry Bragg had described their joint work in not one, but two, papers to one of the most prestigious science journals in the world, *Nature*. In each paper, he gave significant credit to a person he described as "my son".

In the first paper, Henry Bragg wrote, "The rule has been suggested to me as a consequence of an attempt to combine Dr Laue's theory with a fact which my son pointed out to me, viz. that all of the directions of the secondary pencils in this position of the crystal are 'avenues' between the crystal atoms."

In the second paper, the father wrote, "In a paper read recently before the Cambridge Philosophical Society my son has given a theory which makes it possible to calculate the position of the spots for all dispositions of the crystal and photographic plate."

In neither paper did he point out that "my son" was Lawrence Bragg. This oversight was not malicious – but it was still a hurtful oversight. However, this did contribute to "Bragg's lifelong vigilance to see that credit was always given where due – a key element in his successes as a scientific administrator".

33

ANAESTHETICS DOSE

WHEN IT COMES TO PAIN, I AM A TOTAL WEAKLING. I AM NOT EVEN ONE LITTLE BIT STOIC. (HOWEVER, I AM VERY HAPPY THAT PAIN WILL WARN ME OF POTENTIAL DANGERS, SUCH AS A BEE STING WHEN NEAR A HIVE, OR A MILD BURN WHEN NEAR A ROARING FIRE.) BUT I DO NOT THINK THAT PAIN MAKES ME A BETTER PERSON. AND SO, WHEN IT COMES TO ANY KIND OF SURGERY, I TAKE ALL THE ANAESTHETIC OPTIONS I CAN GET.

But how much anaesthetic gas is the right amount to give? Surprisingly, we learnt that by trial and error, and by "carving up" healthy young US Marines – who, unfortunately for their sake, were bound by duty.

WHAT WE STILL DON'T KNOW

Surprisingly, being comfortably numb is an exacting and complicated science. But first, let's dive into a few mysteries.

Even today, we don't fully understand how anaesthetics work, i.e. their mechanism of action. Even so, anaesthetists – or as the Americans laboriously call them, "anaesthesiologists" – know how much to give, and when to give it. (Check out "Mysterious Anaesthetics" in my 22nd book, *Bubbles, Bum Breath and Botox*.) Rest assured, thanks to the long training they receive, specialists know how to give anaesthetics safely.

Of course, the Universe is a Dangerous Place, and everything in life is risky. This means that there are risks in taking anaesthetics. But presumably, if your doctor has recommended you have a procedure requiring anaesthetic, the risk of not having the procedure would be much greater. In a similar fashion, we know that X-Rays carry a small risk of cancer. But you don't get an X-Ray just because you are bored – you get X-Rayed to try to make an essential diagnosis.

Another fundamental mystery is that we still don't know how the brain transitions between Consciousness and Unconsciousness. And we don't know why going down into Unconsciousness follows a different pathway from coming back up into Consciousness.

And we definitely don't understand how the "Anaesthetic Bomb" works. You've probably seen it in the movies. The "Goodies" disable the "Baddies" by rolling an Anaesthetic Bomb across the floor, whereupon it puffs out a Magic Gas – and the Baddies all fall unconscious to the floor, and *all* at the same time. That's right, they each get exactly the dose of gas needed to make them unconscious. Not one of them dies from too much gas, and not one of them remains conscious from too little gas. (Read "Anaesthetic Bomb", in my 23rd book, *Great Mythconceptions*.)

BLEEDING MARINES

Now back to how anaesthetists know how much anaesthetic gas to give.

It's as a result, in part, from experiments on US Marines – who were a bit of a captive audience.

Let me introduce you to the Minimum Alveolar Concentration (MAC). A single MAC (i.e. 1 MAC) is the concentration of an anaesthetic gas inside the lungs that will sedate 50 per cent of the "population", so that they will not move or react to, a "standard surgical incision".

In these experiments, the "population" was healthy young American Marines, and the "standard surgical incision" was a one-centimetre long cut in their forearm. Talking "scientifically", the MAC was derived from this experimental data.

KIDS NEED MORE ANAESTHETIC

It turns out that kids need proportionally greater doses of anaesthetics than adults – relative to their body weight. (This holds true for both gaseous and intravenous anaesthetics.) In other words, weight for weight, kids have a higher MAC. Why?

It's complicated, because normal human physiology is complicated. But we do know a few of the factors involved.

Compared to adults, kids have a proportionally greater cardiac output (the volume of blood the heart pumps in each minute). Their brains, where the anaesthetics work, are proportionally bigger – so they need more "sleeping potion". Their livers, where some of the anaesthetics get broken down, are also proportionally larger.

Kids also have more a more robust cardiovascular system than adults. So they are less compromised by anaesthetics. Adults can easily become hypotensive under anaesthetic – this means that their blood pressure can drop to dangerously low levels.

Another factor is that kids have a different Central Nervous System from adults. We're not sure exactly what the differences are in the Central Nervous System – more synapses, different Blood–Brain

HALOGEN AND ANAESTHETIC?

Halogens are very reactive elements. They are over on the right-hand side of the Periodic Table, right next to the "inert" Noble Gases.

The more halogens that gaseous anaesthetics carry, the less the liver metabolises and removes them. The halogen is usually fluorine, but can also be chlorine and bromine. In general, fluorinated volatile gaseous anaesthetics are *not* removed via the liver, but by the lungs.

Barrier, etc. Time will tell … (Read more about the Blood–Brain Barrier, and its mysteries and potentialities in my 34th book, *Game of Knowns*.)

HOW MUCH MORE?

To make it simple, let's consider just one very commonly used anaesthetic gas, sevoflurane.

You can use anaesthetic gases in premature babies that are born too early – but it's difficult. The trouble is that to achieve an adequate or sufficient Minimum Alveolar Concentration, you have to skirt around the risk of a low blood pressure and a low heart rate. "Premmies" have a lower MAC than babies that were delivered at the regular time, "full-term" babies.

Newborn full-term babies need about 2.0 to 2.4 per cent of sevoflurane in their lungs to achieve 1 MAC. The remaining 98 per cent or so of their lungs is filled with air – often with some extra oxygen added to give a safety margin.

KIDS AND OPIATES

While kids are less sensitive to general anaesthetics, they are more sensitive to opiates.

It seems that their respiratory centres are more sensitive to these drugs – so they are more likely to stop breathing. Also, young children do not metabolise and clear opiates as well as adults do. Overall, opiates may have a more exaggerated and unpredictable effect on children, compared to adults.

The percentage level of sevoflurane needed to cause anaesthesia increases with age until around six months of age – when kids need a level of 2.5–3.0 per cent of anaesthetic gas.

The average kid older than that would need about 2.4 per cent to achieve 1 MAC. The level needed continues to taper off, in a fairly straight-line fashion, until adulthood – when it reaches about 2.1 per cent. It then remains static in the twenties, thirties and forties. Older people, and the unwell, need less anaesthetic to reach 1 MAC.

As an aside, when you're really old enough, you won't need much anaesthetic to knock you out – for good.

34

EXPENSIVE TEACUP

CURRENTLY, THE WORLD'S MOST EXPENSIVE CUP IS A 500-YEAR-OLD CHICKEN CUP, MEASURING JUST 79 MILLIMETRES ACROSS. ACCORDING TO SOTHEBY'S, WHO SOLD IT AT AUCTION, "THE TERM 'CHICKEN CUP' DENOTES A TINY PORCELAIN WINE CUP PAINTED WITH COCKS, HENS AND CHICKS."

OK, we know that the presentation of food or drink affects how it tastes. But if you can spend HK$250 million (about AU$40 million, or GBP£20 million) on a cup, you must really like having a good cuppa. But here's a problem that most of us don't have to worry about – how do you actually physically pay for it when the hammer falls? After all, that's a huge amount of cash to carry around.

CHICKEN CUP 101

Chicken Cups are regarded as the crowning glory of Chinese porcelain. They were made very briefly in the final years of the Chenghua reign (1465–1487). At this point Chinese potters used higher levels of aluminium oxide and lower levels of iron oxide, compared with earlier periods. This allowed higher firing temperatures that produced an extremely fine, very white porcelain that is simultaneously dense and silky. Furthermore, the glaze over the porcelain had low levels of iron and calcium oxide, making its distinctive soft sheen both very clear and fine.

Chicken Cups are outstanding for their remarkable porcelain, their wide range of colours, and a charming, uncontrived and unassuming painting style. There are only some 16 genuine Chicken Cups known. Eight of them are held in the National Palace Museum in Taipei. Another five are held in various other museums, while the remaining three are in private hands. The HK$250 million Chicken Cup is one of those three, and had previously been owned by Swiss pharmaceutical tycoons, the Zuellig brothers.

ECCENTRIC BILLIONAIRE

The self-made billionaire Liu Yiquan, who bought this Chicken Cup on 8 April 2014 at Sotheby's Hong Kong, created immediate controversy. Mr Liu said, "It happened when I was paying. A Sotheby's staffer poured me some tea. I saw the [Chicken Cup] and excitedly poured some of that tea into the cup and drank a little. Such a simple thing, what is so crazy about that?"

Well, is it a wine cup (as the Sotheby's catalogue claims) or is it a teacup? Either way, that didn't stop Mr Liu having a quick cuppa.

Some people were upset that he could have potentially dropped and broken this important part of Chinese history. However, he won them over with the fact he was bringing it back to his art museum in Shanghai – making it the only Chicken Cup in mainland China.

But here's the funny bit. When Mr Liu bought the Chicken Cup at auction, he didn't have GBP£20 million in his back pocket. He did have his black Titanium American Express Centurion card – with no upper limit. But there was still a problem.

Sotheby's has a limit of GBP£900,000 for each card transaction. So Mr Liu had to call on his Confucian-like reserves of patience – and swipe his Centurion AmEx card 24 times.

After the hard yakka of swiping 24 times, I can see why Mr Liu needed a restorative cuppa – immediately.

FREQUENT FLYER MILES

For buying his expensive cup, Mr Liu gained an unexpected bonus of 28 million Frequent Flyer Miles in one hit (or 24 hits, if you separately count each swipe of his Centurion AmEx card). These miles could be immediately converted into all kinds of other rewards, such as US$180,000 of vouchers at the Hong Kong supermarket PARKnSHOP.

He did burn some miles to fly his family to New York and stay at the St Regis Hotel. Mind you, that was after he had spent US$45 million on an embroidered silk thangka – a Tibetan Buddhist painting on fabric.

35

FAT: WHERE DOES IT GO?

WHEN SOMEBODY IS TEMPTED BY A DELICIOUS SLICE OF CAKE, YOU MIGHT HAVE HEARD THEM PARAPHRASE THE WORDS UTTERED BY THE WAIFISH SUPERMODEL KATE MOSS, "NOTHING TASTES AS GOOD AS SKINNY FEELS".

So let's talk about where fat goes.

Where does fat go, when you're putting it on? Surprisingly, it gets laid down differently in your upper and lower body.

And where does fat go, when you're getting rid of it? The atoms leave your body as you would expect – via your major excretory organ. But what is this body part? You might be surprised . . .

FAT IS ESSENTIAL

You can't live without fat.

First, fats are a great high-density source of energy, easily stored inside fat cells. After all, a man was able to live for one year and 17 days without eating any food. He survived purely on his stored fat - all 125 kilograms of it, which he "burnt up" at the rate of about two and a quarter kilograms for each week of his fast. (I discuss this in more detail in the story "Fast for One Year" in my 32nd book, *50 Shades of Grey Matter.*)

Second, fats are a major structural component of the membranes in each and every cell of your body. Furthermore, fats such as prostaglandins and hormone steroids are used as chemical messengers inside your body. The essential vitamins A, D, E, and K are soluble in fat.

And of course, fats act as a shock absorber to protect your internal organs and to insulate you from extremes in temperature. Think of whales and their blubber in the freezing ocean . . .

FAT 101

In the body, fats are usually stored inside fat cells as molecules that have the generic name "triglyceride". There are many different triglycerides. They are made from various combinations of just three different atoms – carbon, hydrogen and oxygen.

A triglyceride molecule looks like the letter E. The vertical stroke, or backbone, of the letter E is a glycerol molecule. The three horizontal strokes of the letter E are three separate molecules called "fatty acids".

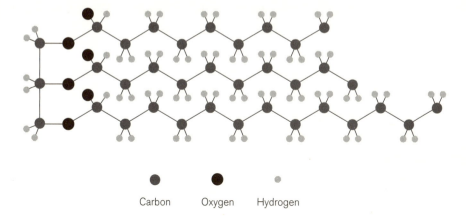

Carbon   Oxygen   Hydrogen

(In this drawing, the glycerol is the vertical-ish bunch of atoms on the left – the three fatty acids are drawn with only 9, 10 and 13 carbon atoms, instead of the average 18 or so).

These three fatty acids can vary a lot. They can all be identical, or they can each be different, or anything in between.

A triglyceride molecule can't enter a cell. But its individual components (glycerol and fatty acid molecules) can.

The "average" triglyceride molecule has about 165 atoms – 55 carbon atoms, 104 hydrogen atoms, and only six oxygen atoms.

ATOMS

Richard Feynman wondered what was the most important single piece of information that we could leave to future generations, in case our society totally vanished. He thought it should be, "Everything is made from atoms."

I still get amazed that the same atoms (carbon, hydrogen, oxygen and nitrogen) make up both our food and the table from which we eat.

FAT ENTERS BODY

We used to think that once you got to about 20 years of age, you never made any more fat cells. But that turned out to be wrong.

One study deliberately overfed 28 healthy, normal-weight men and women. (I know, it's a small sample size.) And yes, the extra food ended up as triglyceride molecules in their fat cells – an extra 3.5 kilograms worth. But where did these atoms end up?

In the upper body, the volunteers laid down 1.9 kilograms of fat. As expected, their fat cells in this area just swelled up and got bigger and plumper.

Surprisingly this did not happen in the lower body, where they added 1.6 kilograms of fat. No, down below the belt, the fat cells stayed the same size. However, the volunteers actually "grew" an extra 2.6 billion new fat cells!

FAT LEAVES BODY

Now that was News – but not as big as the surprising results when two researchers asked where fat goes when you lose it. The researchers were the physicist Ruben Meerman (also known as the Surfing Scientist) from the Australian Broadcasting Corporation, and Professor Andrew Brown from the University of New South Wales.

They carried out a survey. They asked 50 doctors, 50 dieticians and 50 personal trainers the question, "When somebody loses weight, where does it go?" About 60 per cent gave the answer that the fat got converted into heat or energy. This answer was incorrect. The only way that atoms can vanish and turn into heat or energy is via nuclear reactions. The human body doesn't do nuclear reactions – it does only chemical reactions.

Only a few per cent – and they were the dieticians – got the correct answer: the triglyceride molecules get broken down, and then the atoms are rearranged into molecules of carbon dioxide and water (CO_2 and H_2O). In the course of this rearrangment, energy was given off.

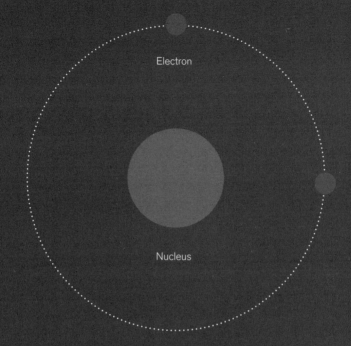

NUCLEAR VS CHEMICAL

As far as atoms are concerned, there are two major pathways involving energy – chemical and nuclear.

In chemical energy, you deal with electrons (the little negative guys in the cloud around the much heavier, positive nucleus).

When molecules get broken down into smaller molecules or individual atoms, those atoms and their electrons get shuffled into new arrangements – and energy can be released. This is how we get energy from chemical burning. The atoms don't vanish – they just get shuffled around into new combinations. The nucleus does not change at all.

But in nuclear energy, the nucleus (the centre, or core, of the atom) gets involved. There are two alternatives. A big nucleus gets split into smaller ones. Or small nuclei get fused into a bigger one.

Thanks to Einstein's famous $E = mc^2$ equation, when atoms get vaporised and turned totally into energy, the numbers are enormous. Vaporising 10 kilograms of fat releases the equivalent of 10,000 Nagasaki atom bombs. If you lost that fat over some 100 days, you would release the energy of 100 atom bombs per day – roughly one every 15 minutes. Your friends couldn't help but notice that were you losing weight . . .

WE ALL CARRY DIFFERENT PERCENTAGES OF FAT

Fat is surprisingly difficult to measure precisely. These figures come from the American Council on Exercise (a private, not a government, body) – so take them with a grain of salt.

There are five categories of fat percentage, ranging from "Essential" to "Obese". "Essential" is the bare minimum needed to

FAT CHEMISTRY

Let's look at what happens when we start (with fat) and burn it (CO_2 and H_2O). Here's the equation, seeing as you desperately wanted it.

$$C_{55}H_{104}O_6 + 78\,O_2 \rightarrow 55\,CO_2 + 52\,H_2O + energy$$

We start with a single triglyceride molecule. The average triglyceride molecule has 55 carbon atoms and 104 hydrogen atoms – but only a very small number of oxygen atoms, just six of them.

We finish with 55 molecules of carbon dioxide, and 52 molecules of water. You can see that carbon dioxide (CO_2) has two atoms of oxygen, while water (H_2O) has one atom of oxygen.

Now here's the important part. We started with only a few atoms of oxygen, but ended up with lots of them. Where did they come from? When your body burns fat, it has to inhale lots and lots of oxygen from the air (78 molecules of them to burn one molecule of triglyceride).

Suppose you want to lose 10 kilograms of fat. That means you're going to have to add 29 kilograms of oxygen – that's a lot of breathing, and huffing and puffing. And to finish the process, you're going to get rid of some 28 kilograms of carbon dioxide and 11 kilograms of water.

And how are you going to get rid of this 39 kilograms of carbon dioxide and water?

survive. You would get unwell if you stayed at, or under this level, for a long time.

Bodybuilders try to adjust their fat levels so that they drop down into the "essential" level for that brief interval when they compete on stage. The purpose of this is to make muscles show up better, thanks to the lower fat levels in the

skin. However, at this low fat level, they sometimes faint when showering.

Women need more fat than men, so they can continue to menstruate cyclically and have children.

	WOMEN (% FAT)	MEN (% FAT)
ESSENTIAL	10–13	2–5
FIT LEAN ATHLETE	14–20	6–13
FIT PERSON	21–24	14–17
AVERAGE	25–31	18–24
OBESE	32–54	25–42

YOUR MAJOR EXCRETORY ORGAN

Overwhelmingly, by breathing them out.

You might lose a tiny amount of this water as sweat or urine. But the vast majority of those atoms that originally made up triglyceride molecules exit your mouth and nose as carbon dioxide and water. However, you have to do a lot of breathing – each breath removes only 33 milligrams of carbon dioxide.

This means that your major excretory organ is your lungs. And your lungs send the water and carbon dioxide out via your nose and mouth.

It is true that you are what you eat. And you can't outrun a muffin – in one minute, you can eat more than you can exercise off in an hour.

But can you unlock the carbon in your fat cells, and lose weight, by breathing more? Kind of . . . The most effective way to breathe more often is to exercise. So, eat less and move more.

36

EARTH'S SPIN AXIS IS SHIFTING

WE ARE ONLY A LITTLE WAY INTO THE 21ST CENTURY, AND THE SIGNS OF GLOBAL WARMING ARE CLEAR. OBVIOUSLY, THERE ARE THE RELENTLESSLY INCREASING TEMPERATURES OF THE AIR, LAND AND THE OCEANS. AND THERE ARE MORE AND MORE DROUGHTS IN EAST AFRICA, STRANDED POLAR BEARS IN THE ARCTIC, BLEACHED CORAL REEFS IN THE TROPICS, AND RETREATING GLACIERS ON LAND.

Even so, the latest finding really surprised me. By burning huge quantities of fossil fuels, we humans have actually tipped the Earth off its axis by a tiny amount.

SPIN AXIS

Let me emphasise how tiny the tipping is. Each year since 2005, our Global Warming activities have shifted the Spin Axis from its previous path by centimetres – not kilometres.

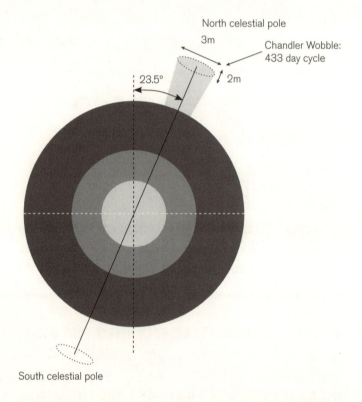

The North–South Spin Axis of the Earth runs, of course, through the North Geographical Pole in the Arctic Ocean, and the South Geographical Pole in the Antarctic. (I'm not talking about the North and South Magnetic Poles, just the Geographical Poles.)

DIRT

We humans now shift as much dirt as all the rivers on Earth added together. It comes to about 20 billion tonnes per year – mostly coal, followed by iron ore.

This is a huge amount. But by using carbon dioxide, we moved an amount of water 30 times greater. We melted ice on land to water, which then shifted all over the entire planet.

WOBBLING POLE

However, as the Earth spins on its own axis, the position of the North Pole is not dead true – it wobbles a little. There's a bunch of reasons.

The Earth is not perfectly spherical. Instead, like Tweedledum and Tweedledee from *Through the Looking Glass*, it's a bit flattened at the Poles, and a bit bulging at the Equator.

In addition, the surface is not smooth – it's pretty bumpy. Mountains poke up towards space, while oceans dip down into the solid crust.

Furthermore, our planet is not perfectly rigid, but somewhat elastic. Yes, it does have a solid crust at the surface – but it's very thin. The Earth is made mostly of molten rock and then liquid iron, with a core of solid iron. So even today, parts of the crust that carried heavy ice sheets 20,000 years ago are still slowly rising (an effect known as the "isostatic rebound" or "post-glacial rebound").

As a result of these (and other) factors, when the Earth rotates on its own axis over the course of a day, that spin axis wobbles a little. There are lots of individual wobbles.

A major one is the so-called Chandler Wobble, which the American astronomer Seth Carlo Chandler discovered in 1891. Over a period of slightly more than a year (about 433 days), the Chandler Wobble shifts the North Pole over a rough circle or ellipse – about several metres across.

About two thirds of the Chandler Wobble seems to be caused by ocean currents, and about one third by winds in the Earth's atmosphere.

MEASURING POLE POSITION

We've been measuring the position of the North Pole for over a century – since 1899.

However, early efforts were fairly crude. More accurate methods arrived in the early 1980s – Very Long Baseline Interferometry, Satellite Laser Ranging, Lunar Laser Ranging. These were at least 100 times better than earlier measurements.

From 1982 to 2005, we measured the location of the North Pole as drifting slowly South towards Labrador, around six to seven centimetres each year. (This is after removing the various wobbles, and working with the average location of the North Pole.) But in 2005, the motions of the North Pole suddenly, and without any warning, flipped in three unexpected ways.

First, with regard to direction, the North Pole chucked a leftie and started heading East, parallel to the Equator. It's still heading East.

Second, with regard to motion, the North Pole more than tripled its drift speed to about 24 or so centimetres per year. It's still drifting at this higher speed.

Third, the Chandler Wobble changed phase. So far, our scientists do not have a good answer as to why the Chandler Wobble changed phase.

However, they do have a good answer for the tipping of the Spin Axis.

HOW WE TIPPED THE EARTH

It's plain and simple. Rapid melting of ice on land has made the drift velocity of the North Pole accelerate, and has changed its direction of travel to the East.

This solid ice is now liquid water spread everywhere across the planet. We know where the ice was, we know where it's gone to – and the maths all fits in with the observed changes to motions of the North Pole.

Since the early 1900s, we've used satellites to take many tens of millions of accurate measurements of these land ice changes. These

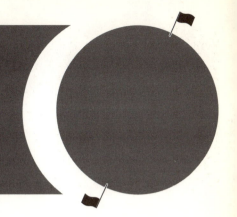

Since 2002, a pair of Gravity Recovery and Climate Experiment (GRACE) satellites have improved the precision and accuracy of the data available. GRACE measures how the gravity on different parts of the planet changes over time, following variations in polar ice sheets, mountain glaciers, movements of water on land, the atmosphere and the oceans.

measurement continue to this day, with the European Space Agency's CryoSat-2 (launched in 2010).

Recent analysis tells us that between 2011 and 2014, Greenland, Antarctica and mountain glaciers were losing about 600 billion tonnes of land ice per year. Most of this ice came from Greenland. This was an increase of two to three times over the loss rate between 2003 and 2009.

This is actually quite astonishing.

It's hard to imagine something so small as us humans being able to shift something as massive as our whole planet. (Given all the humans on Earth would easily fit into a cubical box just one kilometre on its side.) But we used Global Warming as a Force Multiplier.

We dumped billions of tonnes of carbon dioxide into the atmosphere, which then heated the atmosphere and the oceans. The combination of hotter atmosphere and ocean water then melted over half a trillion tonnes of ice, which then flowed as liquid water into the oceans. This redistribution of water shifted the North–South Spin Axis.

Why did both the Chandler Wobble and the Spin Axis shift *suddenly* – instead of *gradually*? We don't why – yet.

Perhaps it's like pushing a pencil towards the edge of a table. You push and you push and you push, and it's still on the table. But then you give it just one more tiny push. Its Centre of Gravity is no longer supported by the table, and it suddenly falls to the floor.

So if we push and push at the Balance of our Planet, it may well respond by throwing a real wobbly of its own...

37

BOUNCING BATTERIES

AN ORDINARY FAMILY CAN EASILY HAVE A DOZEN OR MORE ALKALINE BATTERIES IN VARIOUS DEVICES AROUND THE HOUSE. THEY'LL BE IN SMOKE DETECTORS AND REMOTE CONTROLS, THERMOSTAT CONTROLS AND WALL CLOCKS, CHILDREN'S TOYS AND WEATHER STATIONS, SOLAR CELL MONITORS AND FLASHLIGHTS, AND THE LIKE.

BATTERIES ...

Can be either "acid" (like the one in your car), or "alkaline".

Alkaline wet batteries were invented independently by Waldemar Jungner in 1899, and Thomas Edison in 1901. In the 1950s, Lewis Urry invented the dry zinc–manganese dioxide alkaline battery. Each year, 10 billion of these alkaline batteries get made and sold. They make up about 30 per cent of the batteries sold around the globe. In 2013, the worldwide sales of AA alkaline batteries generated US$1.8 billion.

The advantages of alkaline batteries include being able to operate at low temperatures, and to deliver high

But how do you check how full a battery is? Sure, an official battery tester is one way. But it's much more fun to see how high the battery bounces when you drop it. Within limits, the less it bounces, the more charge it carries.

This is because electrical changes inside the battery cause chemical changes, which then cause physical changes. As the battery loses charge, a gel-like substance inside turns into a porous solid – and that makes the battery more bouncy.

YOUTUBE VIDEO TO PEER-REVIEWED PAPER

One day when I was quietly minding my own business, I saw a viral YouTube video posted by someone called "bajarider100". It claimed that you could test the State of Charge of a battery by dropping it – and seeing how high it bounces. Wow!

I was skeptical, so I tried it out at home with a half-a-dozen random AA batteries that were lying around the house. I don't have a dedicated Battery Tester, so I measured their State of Charge by shorting them out with a multimeter set to the 10 Amp scale. I measured their "bounciness" by dropping them all from the same height, and seeing how high they jumped up.

currents for relatively long times.
They also have a long shelf life –
up to four years at 20°C.
This makes them suitable for low
electrical drain "fit-and-forget"
uses, such as in smoke detectors.

Their big disadvantage is that most
of them are not rechargeable.
All you get is one use – and then
you have to toss them.

In my simple (but very flawed) experiment, I found no link between the bounciness of a battery and its State Of Charge. But there *is* a real link. The woeful experimental technique was entirely 'my bad' – and that was why I didn't see the link.

First, the batteries came from different manufacturers. This made them difficult to compare directly. Second, I neglected to ensure that they landed truly vertically (for example, by running them through a long hollow tube). And finally – and obviously – my sample size was too small.

The idea that there might be something to this "crazy" claim kept niggling at my brain. I needed some hard information on this. By "hard", I mean something more than a 90-second video made by a random person and posted on the interwebs.

So I went looking and I found the Mother Lode – a paper in the *Journal of Materials Chemistry A* (a proper peer-reviewed journal). The paper was written by authentic engineers and real scientists. It turns out that under certain circumstances, there is a link between "how flat a battery is" and "how far the battery bounces back up when dropped".

SCIENTIFIC TRUTH

This is what the journal told me. When the battery is between 80 and 100 per cent full, it will bounce back upwards about 23 per cent of the distance you dropped it.

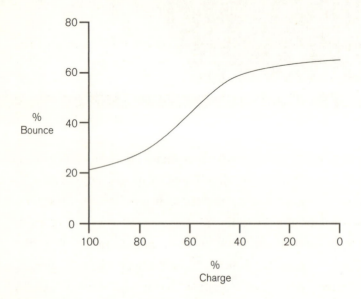

And then, as you gradually flatten the battery from 80 per cent full down to 50 per cent, it will bounce higher and higher, getting up to about 60 per cent of the distance you dropped it. But once the battery is down to around half full, you've pretty well reached maximum bounce. Even if you discharge it down to totally empty, you don't get a lot more bounce – perhaps just another six per cent or so.

But, in between 50 and 80 per cent full, the flatter the battery – the higher it bounces.

How come?

POSITIVE OR NEGATIVE?

In "regular" batteries (AAA, AA, C, D) the usual convention is that the positive terminal is the one with the bump. That leaves the flattened end to be the negative terminal.

It's usually the other way around in the smaller "button" batteries as found in watches. Normally, the flat end is the positive. I don't know why the convention is different.

ALKALINE AA BATTERY 101

To understand why, you need to know the anatomy of a typical alkaline AA battery.

The first thing you see is the plastic skin, imprinted with the brand name, some necessary information (size, place of manufacture etc), a warning to not eat the battery, an expiry date, and so on. The plastic is tightly wrapped around a cylindrical steel case. Inside that case are three concentric layers.

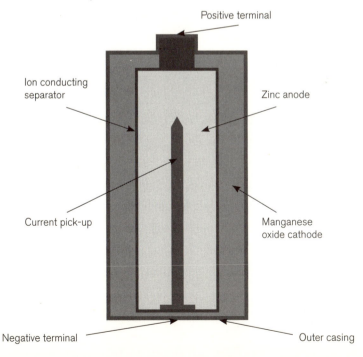

TESTING ...

It's good to have a battery tester at home. Let me tell you why you should test batteries.

We have many devices that use more than one battery. When the device stops working, I first test all the batteries. More often than not, only one of the batteries has gone flat, leaving the others at three-quarter charge. This means that I can save money and reduce waste by swapping out only the single dead battery.

The disadvantage is that, for a while, I would keep using the batteries until they were dead flat. Unfortunately, sometimes a single battery would corrode and then "kill" the device (usually a TV/audio/PRV/etc remote control) by leaking

---

The outermost layer is the cathode, made from a paste of manganese dioxide. It's a few millimetres thick. (It includes some carbon powder, purely to help it carry electricity better.) It's connected directly to the positive end of the battery – the bumpy bit, at the top.

Heading inwards, the next layer is a thin-walled non-metallic cylinder called the separator. It's usually made from some kind of porous fabric, such as cellulose, or a synthetic polymer. It keeps the outer manganese oxide cathode physically separated from the gelatinous zinc anode inside it. But it still lets electrons pass through.

And right at the centre is the zinc gel anode. The gel carries about two and a half grams of tiny zinc particles. The zinc anode is connected directly to the negative end of the battery, at the bottom. It does this via a thin metal rod, known as the "Current Pick-Up".

This gel also contains some potassium hydroxide, which provides the hydrogen and oxygen that is needed to make the chemistry go through its reactions.

nasty chemicals over delicate electronics. So now, I do it differently.

I use the battery tester *and* I check all the batteries in the house for corrosion at the beginning of each season (i.e., every three months). I still get caught occasionally. The single battery in my noise-cancelling headphones corroded while still having a good charge.

Luckily, the corrosion hadn't reached and destroyed any electronics, and I was able to clean the contaminated area with cotton buds soaked in methylated spirits, and a can of compressed air.

Warning! The chemicals coming out of corroded batteries are nasty. Keep them off your skin and always wash your hands.

PHYSICAL-ELECTRO-CHEMISTRY

So when you connect the positive and negative ends of the battery to a flashlight bulb, you complete an electrical circuit – and chemical reactions start happening. These reactions liberate electrons. A little electrical miracle happens – but what about physical events?

There are physical changes –and they happen inside the zinc particle gel that runs along the centre line of the battery. (Just ignore the changes to the manganese dioxide cathode.)

In the first stage, while the battery is still close to fully charged, there are chemical changes, but no major physical changes. As the battery drops from fully charged to 80 per cent charged, the zinc atoms get converted into individual molecules of zinc hydroxide. They gradually saturate the gel of the tiny zinc particles. By the time the battery gets down to 80 per cent charged, the gel is fully saturated with zinc hydroxide. Over this range from 100 per cent down to 80 per cent charged, the gel is still very jelly-like and flexible – it really doesn't change much at all. Because the gel doesn't change much, neither does the bounciness of the battery.

In a gel, the particles can all move easily past each other. In a battery that's between 100 per cent and 80 per cent charged, the soft gloopy zinc gel absorbs the energy of the impact when the battery is dropped – and turns most of it into heat. This leaves hardly any energy to bounce the battery up off the table.

In the second stage, more dramatic physical changes happen as you gradually flatten the battery down from 80 to 50 per cent full. The zinc hydroxide gel converts into a stiffer chemical called zinc oxide – more like wood than a jelly. The zinc oxide forms tiny bridges with other zinc oxide particles. This change begins on the outside (at the separator) and slowly migrates inward. The more zinc oxide there is, the more bouncy the battery becomes.

The extra bounce is because the energy of the impact mostly passes through the stiff zinc oxide without being absorbed. The impact energy then hits the inside of the top of the battery, reflects back down to the bottom – and propels the whole battery into the air.

(It might be a coincidence, but zinc oxide is incorporated into golf balls to add extra bounce.)

But once you get to 50 per cent charge, the zinc anode is as stiff as it's going to get. Adding more zinc oxide does not make it any stiffer. So you don't get more bounce after that point.

TAKE-HOME MESSAGE

The bounciness of a battery can roughly (and excitingly) tell you how much charge it carries. However, it's not a linear relationship between "bounce" and "charge".

When the battery is pretty full (80–100 per cent), you get a pretty constant 23 per cent bounce.

In the relatively narrow range between 50 and 80 per cent charged, the more empty the battery, the better it bounces. By the time you get to half full, you get about 60 per cent bounce.

From 50 per cent fully charged down to dead flat, you get a pretty constant bounce – climbing only slightly, to around 66 per cent. Now, a battery that's half-full is usable, but a flat battery is not. So in this lower range, the bounce test is not much help.

From a philosophical point of view, batteries should never be taken for granted. They're a little power cell of wonder that can give unexpected bounce to your life ...

ALKALINE CHEMISTRY

When you join the positive and negative ends of the battery to a "load" (flashlight, alarm clock, etc.), stuff happens. Inside the jelly-like anode, a few chemical reactions start. Importantly for a battery, electrons are given off in these reactions.

$$Zn + 4OH^- \rightarrow Zn(OH)_4^{2-} + 2e^-$$

$$Zn(OH)_4^{2-} \rightarrow ZnO + H_2O + 2OH^-$$

The cathode uses the electrons, after they have passed through the "load". This starts a different chemical reaction.

$$MnO_2 + H_2O + e^- \rightarrow MnOOH + OH^-$$

So as the battery discharges, zinc turns into zinc oxide, and manganese dioxide turns into manganese oxyhydroxide.

More importantly with regard to making usable electricity, electrons are released from the zinc, and before they are collected by the manganese dioxide, travel through the "load". As a result, a flashlight can shine, an alarm clock can buzz, and so on.

REFERENCES

FOR FURTHER READING ... READ ON!

01 RADIOACTIVE YOU

Living With Risk, by The British Medical Association, John Wiley & Sons, London, 1987.

"Life Is Rad", by Stephen Cass and Corinna Wu, *Discover*, June 2007, Vol. 28, No. 6, page 76.

02 SLIPPERY BANANA PEELS

"Frictional Coefficient under Banana Skin", by Kiyoshi Mabuchi et al., *Tribology Online*, Vol. 7, No. 3, 30 September 2012, pages 147–151.

03 BREATHE IN SPACE

"Status of the International Space Station Regenerative ECLSS Water Recovery and Oxygen Generation Systems", by Robert M. Bagdigian and Dale Cloud, NASA, 2005, http://archive.org/details/nasa_techdoc_20050207456.

"International Space Station: Environmental Control and Life Support System", NASA Facts, http://www.nasa.gov/sites/default/files/104840main_eclss.pdf.

04 A HOLE THROUGH THE EARTH

"De Paris à Rio de Janeiro en 42'11''", by A. Redier, *La Nature*, 19 May 1883, No. 520.

"Through the Earth in Forty Minutes", by Paul W. Cooper, *American Journal of Physics*, January 1966, Vol. 34, No. 1, pages 68–70.

"An Example of the Need for Adequate References", by Philip G. Kirmser, *American Journal of Physics*, August 1966, Vol. 34, No. 8, page 701.

"Further Commentary on 'Through the Earth in Forty minutes'", by Paul W. Cooper, *American Journal of Physics*, August 1966, Vol. 34, No. 8, pages 703–704.

"The Gravity Tunnel in a Non-Uniform Earth", by Alexander R. Klotz, *American Journal of Physics*, March 2015, Vol. 83, No. 3, pages 231–237.

05 QUANTUM LIFE

"The Quantum Life", by Paul Davies, *Physics World*, July 2009, Vol. 22, No. 7, pages 24–28.

"The Dawn of Quantum Biology", by Philip Ball, *Nature*, 16 June 2011, Vol. 474, No. 7351, pages 272–274.

"Quantum Life: The Weirdness Inside Us", by Michael Brooks, *New Scientist*, 1 October 2011, Vol. 211, No. 2832, pages 34–37.

"Engineering Coherence Among Excited States in Synthetic Heterodimer Systems", by Dugan Hayes et al., *Science*, 21 June 2013, Vol. 340, No. 6139, pages 1431–1434.

"Quantum Coherent Energy Transfer over Varying Pathways in Single Light-Harvesting Complexes", by Richard Hildner et al., *Science*, 21 June 2013, Vol. 340, No. 6139, pages 1448–1451.

Life on the Edge: The Coming of Age of Quantum Biology, by Jim Al-Khalili and Johnjoe McFadden, Bantam Press, 2014.

"Quantum Mechanics Boosts Photosynthesis", by Edwin Cartlidge, 4 February 2014, *Physics World*, http://physicsworld.com/cws/article/news/2010/feb/04/quantum-mechanics-boosts-photosynthesis

"You're Powered by Quantum Mechanics. No, Really . . .", by Jim Al-Khalili and Johnjoe McFadden, *The Guardian*, 26 October 2014, http://www.theguardian.com/science/2014/oct/26/youre-powered-by-quantum-mechanics-biology

"Solving Biology's Mysteries Using Quantum Mechanics", by Zeeya Merali, *Discover*, December 2014, pages 44–49.

"Quantum Measurement Is for the Birds, But is Not Essential for Plants", by Hamish Johnston, 6 March 2015, http://physicsworld.com/cws/article/news/2015/mar/06/quantum-measurement-is-for-the-birds-but-is-not-essential-for-plants

06 NAKED MOLE RATS DON'T GET CANCER?

"Eusociality in a Mammal: Cooperative Breeding in Naked Mole-Rat Colonies", by J.U.M. Jarvis, *Science*, 1 May 1981, Vol. 213, No. 4494, pages 571–573.

"Societies from Underground", by Gail R. Michener, *Science*, 16 August 1991, Vol. 253, No. 5021, pages 803–804.

"Late Pliocene Faunal Turnover in the Turkana Basin, Kenya and Ethiopa", by Anna K. Behrensmeyer et al., *Science*, 28 November 1997, Vol. 278, No. 5343, pages 1589–1594.

"Hypersensitivity to Contact Inhibition Provides a Clue to Cancer Resistance of Naked Mole-Rat", by Andrei Seluanov et al., *PNAS*, 17 November 2009, Vol. 106, No. 46, pages 19352–19357.

"Naked and Ugly: The New Face of Lab Rats", by Kirsten Weir, *New Scientist*, 28 October 2010, Vol. 208, No. 2783, pages 44–47.

"Feeling No Pain", *Science*, 16 December 2011, Vol. 334, No. 6062, page 1471.

"Zoologger: The Longest Tunnels Dug by a Mammal", by Michael Marshall, *New Scientist*, 10 August 2012, http://www.newscientist.com/article/dn22164-zoologger-the-longest-tunnels-dug-by-a-mammal.html

"Cancer Resistance in the Blind Mole Rat is Mediated by Concerted Necrotic Cell Death Mechanism", by Vera Gorbunova et al., *PNAS*, 20 November 2012, Vol. 109, No. 47, pages 19392–19396.

"High-Molecular-Mass Hyaluronan Mediates the Cancer Resistance of the Naked Mole Rat", by Xiao Tian et al., *Nature*, 18 July 2013, Vol. 499, No. 7458, pages 346–349.

"Naked Mole-Rat Has Increased Translational Fidelity Compared with the Mouse, as well as a Unique 28S Ribosomal RNA Cleavage", by Jorge Azpurua et al., *PNAS*, 22 October 2013, Vol. 110, No. 43, pages 17350–17355.

"Natural Selection and Pain Meet at a Sodium Channel", by Gary R. Lewin, *Science*, 25 October 2013, Vol. 342, No. 6157, pages 428–429.

07 STINKY SYNTHETIC SHIRTS

"Microbial Odor Profile of Polyester and Cotton Clothes after a Fitness Session", by Chris Callewaert et al., *Applied and Environmental Microbiology*, November 2014, Vol. 80, No. 21, pages 6611–6619.

08 BLUE DRESS ILLUSION

"The Science of Why No One Agrees On The Colour of This Dress", by Adam Rogers, *Wired*, 26 February 2015, http://www.wired.com/2015/02/science-one-agrees-color-dress.

"The White and Gold (No, Blue and Black!) Dress That Melted the Internet", by Jonathan Mahler, *The New York Times*, 27 February 2015, http://www.nytimes.com/2015/02/28/business/a-simple-question-about-a-dress-and-the-world-weighs-in.html.

"The Inside Story of the 'White Dress, Blue Dress' Drama that Divided a Planet", by Terrence McCoy, *The Washington Post* Morning Mix blog, 27 February 2015, http://www.washingtonpost.com/news/morning-mix/wp/2015/02/27/the-in-side-story-of-the-white-dress-blue-dress-drama-that-divided-a-nation.

"Why Julianne Moore and Taylor Swift See That Dress Differently", by Stephen L. Macknik, *Scientific American* Illusion Chasers blog, 27 February 2015, http://blogs.scientificamerican.com/illusion-chasers/2015/02/27/thatdress.

09 BRONTOSAURUS IS BACK

"Appendix. Art. XI – *Principal Characters of the Coryphodontidae*", by O.C. Marsh, *American Journal of Science*, July 1877, Series 3, Vol. 14, No. 79, pages 81–85.

"Appendix. Art. LIII – Principal Characters of the American Jurassic Dinosaurs, Part V", by O.C. Marsh, *American Journal of Science*, May 1881, Series 3, Vol. 21, No. 125, pages 417–423.

"Supersonic Sauropods? Tail Dynamics in the Diplodocids", by Nathan P. Myhrvold and Philip J. Currie, *Paleobiology*, October 1997, Vol. 23, No. 4, pages 393–409.

"Dinosaurs in Motion", by Carl Zimmer, *Discover*, November 1997, pages 96–109.

"A Specimen-Level Phylogenetic Analysis and Taxonomic Revision of Diplodoci-dae (Dinosauria, Sauropoda)", by Emanuel Tschopp et al., *PeerJ*, 7 April 2015, Vol. 3, e857.

"A Prehistoric Giant Is Revived, If Only in Name", by James Gorman, *The New York Times*, 8 April 2015.

10 STARS ARE POINTY

"An Estimate of the Size and Shape of the Human Lens Fibre in Vivo", by Nicholas A.P. Brown et al., *British Journal of Ophthalmology*, December 1987, Vol. 71, No. 12, pages 916–922.

"The Pulling, Pushing and Fusing of Lens Fibers: A Role for Rho GTPases", by P. Vasantha Rao, *Cell Adhesion and Migration*, July–September 2008, Vol. 2, No. 3, pages 170–173.

"The Surprisingly Complicated Reason Why Stars Look Like They Have Points", by Colin Schultz, *Smithsonian*, 2 September 2014, http://www.smithsonian mag.com/smart-news/surprisingly-complicated-reason-why-stars-seem-have-points-180952587.

11 CORN PASSES STRAIGHT THROUGH

"Why Does the Human Body Not Digest Corn?", by Tina M. St. John, *LIVESTRONG*, 18 December 2013, http://www.livestrong.com/article/477602-why-does-the-human-body-not-digest-corn.

12 NAKED IN THE ANTARCTIC

"The Dome is Home – South Pole History 1975-1990", http://www.southpolestation.com/trivia/history/history.html.

"300 Club time", http://www.southpolestation.com/winter/300club.html.

"Darryn's Antarctic Diary: Week 37, 300 Club – Sunday 25th June – 2nd July 2000", by Darryn Schneider, http://antarctica.kulgun.net/SouthPole/Diary2000/week_37_jul02_2000.shtml.

"South Pole's 300 Club Not For the Cold-Blooded", by Deborah Zabarenko, 13 December, 2006, http://uk.reuters.com/article/2006/12/13/oukoe-uk-antarctica-300club-idUKN1341483220061213.

"There's Exclusive Clubs, then there's the South Pole's '300 Club'", by Scott Sistek, http://www.komonews.com/weather/blogs/scott/93720919.html.

"On Getting Naked in Antarctica" by Svati Kirsten Narula, The Atlantic, 7 January 2014, http://www.theatlantic.com/international/archive/2014/01/on-getting-naked-in-antarctica/282883.

13 THE KILLER DRILLER

"Who Discovered Bacteriophage", by Donna H. Duckworth, *Bacteriological Reviews*, December 1976, Vol. 40, No. 4, pages 793–802.

"Set a Bug to Catch a Bug", by John MacGregor, *New Scientist*, 5 April 2003, http://www.newscientist.com/article/mg17823895-100-set-a-bug-to-catch-a-bug/

"Virus Cleans Up Food Poisoning Bug", by James Randerson, *New Scientist*, 23 April 2003, http://www.newscientist.com/article/dn3652-virus-cleans-up-food-poisoning-bug.

"Bug Killers", by Thomas Häusler, *Nature Medicine*, 1 June 2006, Vol. 12, No. 6, http://www.nature.com/nm/journal/v12/n6/abs/nm0606-600.html.

"The Gene Weavers", by Garry Hamilton, *Nature*, 8 June 2006, Vol. 441, No. 7904, pages 683–685.

"Viruses to Be Sprayed On Cold Meats", *New Scientist*, 27 August 2006, https://www.newscientist.com/article/dn9863-viruses-to-be-sprayed-on-cold-meats.

"Corynebacterium diphtheria: Genome Diversity, Population Structure and Genotyping Perspectives", by Igor Mokrousov, *Infection, Genetics and Evolution*, January 2009, Vol. 9, No. 1, pages 1–15.

"Bacteriophage Therapy: Exploiting Smaller Fleas", by Stan Deresinski, *Clinical Infectious Diseases*, 15 April 2009, Vol. 48, No. 8, pages 1096–1101.

"Viruses Could Kill Superbugs that Antibiotics Can't", by Catherine de Lange, *New Scientist*, 25 April 2009, http://www.newscientist.com/article/mg20227054-500-viruses-could-kill-superbugs-that-antibiotics-cant/

"Fabricating Genetically Engineered High-Power Lithium-Ion Batteries Using Multiple Virus Genes", by Yun Jung Lee et al., *Science*, 22 May 2009, Vol. 324, No. 5930, pages 1051–1055.

"Driller Killers: Turning Bacteria's Weapons on Them", by Roberta Kwok, *New Scientist*, 25 January 2012, http://www.newscientist.com/article/mg21328482-000-driller-killers-turning-bacterias-weapons-on-them.

"Phage Therapy: Concept to Cure", by Eric C. Keen, *Frontiers in Microbiology*, 19 July 2012, http://journal.frontiersin.org/article/10.3389/fmicb.2012.00238/full.

"Rise of the Nano Machines", by Michail C. Roco, *Scientific American*, May 2015, Vol. 308, No. 5, pages 34–35.

"Meet Your New Symbionts: Trillions of Viruses", by Carl Zimmer, *National Geographic* Phenomena blog, 20 May 2013, http://phenomena.nationalgeographic.com/2013/05/20/meet-your-new-symbionts-several-trillion-viruses.

"Bacteriophage Adhering to Mucus Provide a Non–Host-Derived Immunity", by Jeremy J. Barr et al., *PNAS*, 25 June 2013, Vol. 110, No. 26, pages 10771–10776.

"I'm Breeding Biodegradable Batteries from Viruses", *New Scientist*, 26 February 2014, http://www.newscientist.com/article/mg22129580.400-im-breeding-biodegradable-batteries-from-viruses.

"Infections Infected", by Linda Marsa, *Discover*, March 2014, pages 20–22.

"The Age of the Phage", by Shigenobu Matsuzaki et al., *Nature*, 1 May 2014, Vol. 509, No. 7948, page S9.

"Phage Therapy Gets Revitalized", by Sara Reardon, *Nature*, 4 June 2014, Vol. 510, No. 7503, pages 15–16.

"Elizabeth Taylor, My Great-Grandpa, and the Future of Antibiotics", by Michael White, *Pacific Standard*, 27 June 2014, http://www.psmag.com/nature-and-technology/elizabeth-taylor-great-grandpa-future-antibiotics-84626.

"The CD27L and CTP1L Endolysins Targeting *Clostridia* Contain a Built-in Trigger and Release Factor", by Matthew Dunne et al., *PLoS Pathogens*, 24 July 2014, e1004228.

"How to Develop New Antibiotics", by Ezekiel J. Emanuel, *The New York Times*, 24 February 2015, http://www.nytimes.com/2015/02/24/opinion/how-to-develop-new-antibiotics.html.

14 SURPRISING SUN

"Coronal Rain as a Marker for Coronal Heating Mechanisms", by P. Antolin and K. Shibata, 13 October 2009, http://arxiv.org/pdf/0910.2383.pdf.

"Sun's Rain Could Explain Why Corona Heat Is Insane", *New Scientist*, 21 October 2009, http://www.newscientist.com/article/mg20427315–300-suns-rain-could-explain-why-corona-heat-is-insane.

"The Birth Environment of the Solar System", by Fred C. Adams, 29 January 2010, http://arxiv.org/pdf/1001.5444.pdf.

"Searching for the Sun's Long-Lost Siblings", by Ken Croswell, *New Scientist*, 3 April 2012, http://www.newscientist.com/article/mg21428592–200-searching-for-the-suns-long-lost-siblings.

"Elemental Abundances of Solar Sibling Candidates" by I. Ramirez et al., *The Astrophysical Journal*, June 2014, Vol. 787, No. 2, pages 154–170.

"Strangest Star: 6 Things We Didn't Know About the Sun", by Rebecca Boyle, *New Scientist*, 17 September 2014, https://www.newscientist.com/article/mg22329870.500-strangest-star-6-things-we-didnt-know-about-the-sun.

16 COCONUT OIL AND WATER

"Sugar Content of Popular Sweetened Beverages Based on Objective Laboratory Analysis: Focus on Fructose Content", by Emily E. Ventura et al., *Obesity*, 14 October 2010, pages 1–7 (Obesity (2010) doi:10.1038/oby.2010.255).

"Victoria's Secret? Coconut Oil . . . Sales Boom as Model Miranda Kerr Reveals Daily Dose of 'Healthy Fat' Is Key to Her Beauty", by Tamara Cohen, *Daily Mail*, 25 August 2011, http://www.dailymail.co.uk/femail/article-2029573/Victorias-Secret-supermodel-Miranda-Kerrs-coconut-oil-beauty-secret.html.

"Comparison of Coconut Water and a Carbohydrate-Electrolyte Sport Drink on Measures of Hydration and Physical Performance in Exercise-Trained Men", by Douglas S. Kalman et al., *Journal of the International Society of Sports Nutrition*, 2012, Vol. 9, No. 1, pages 1–10.

"Is Coconut Water Healthy?", by Kate Browne, *Choice*, 21 August 2012, http://www.choice.com.au/food-and-drink/drinks/juices-and-smoothies/articles/is-coconut-water-healthy.

"Product Review: Coconut Waters Review – Tests of O.N.E., Vita Coco, and Zico", 13 September 2012, *Consumer Lab*, https://www.consumerlab.com/reviews/coconut-water-one-vita-coco-zico-review/coconut-water/.

"Coconut Water: Miracle Drink or Health Fad?", by Cassie White, *ABC Health and Wellbeing*, 23 July 2013, http://www.abc.net.au/health/thepulse/stories/2013/07/23/3808325.htm.

"Is Coconut Oil Healthy?", by Kate Browne, *Choice*, 23 February 2014, https://www.choice.com.au/food-and-drink/nuts-and-oils/oils/articles/is-coconut-oil-healthy.

17 OLDEST KNOWN COMPLAINT LETTER

Letters from Mesopotamia: Official, Business, and Private Letters on Clay Tablets from Two Millennia, by A. Leo Oppenheim, University of Chicago Press, Chicago and London, 1967.

"Humble Clay Tablets Are Greatest Loss to Science", by Bob Holmes and James Randerson, *New Scientist*, 10 May 2003, Vol. 178, No. 2394, page 8.

"Ancient Customer-Feedback Technology Lasts Millennia", by Chris Baraniuk, *New Scientist*, 2 March 2015, https://www.newscientist.com/article/dn27063-ancient-customer-feedback-technology-lasts-millennia.

18 BREAKING THE SEAL

"The Science of 'Breaking the Seal'", by Matt Soniak, *Mental Floss*, 7 August 2012, http://mentalfloss.com/article/31408/science-breaking-seal.

"A Happy Hour Urban Myth: Is 'Breaking the Seal' a Real Thing", *Body Odd*, NBC News, 18 January 2013, http://bodyodd.nbcnews.com/_news/2013/01/18/16572994-a-happy-hour-urban-myth-is-breaking-the-seal-a-real-thing.

19 WINE GLASS SHAPE AFFECTS FLAVOUR

"Camera Turned Wine Connoisseur", by Jennifer Newton, *ChemistryWorld*, 10 April 2015, http://www.rsc.org/chemistryworld/2015/04/ethanol-escape-wine-glass-camera.

"Wine Snobs Are Right: Glass Shape Does Affect Flavor", by Jennifer Newton and ChemistryWorld, *Scientific American*, 14 April 2015, http://www.scientificamerican.com/article/wine-snobs-are-right-glass-shape-does-affect-flavor.

"A Sniffer-Camera for Imaging of Ethanol Vaporisation from Wine: The Effect of Wine Glass Shape", by Takahiro Arakawa et al., *Analyst*, March 2015, Vol. 140, No. 8, pages 2881–2887.

20 MICROMORT
"Britain's Most Dangerous Roads", by Tobias Jolly, *BBC Focus Magazine*, March 2015, pages 64–71.

21 GREAT BARRIER REEF
"Year of the Reef", by Donald Kennedy, *Science*, 14 December 2007, Vol. 318, No. 5857, page 1695.

"Coral Reefs under Rapid Climate Change and Ocean Acidification", by O. Hoegh-Guldberg et al., *Science*, 14 December 2007, Vol. 318, No. 5857, pages 1737–1742.

"Calcification Rates Drop in Australian Reefs", by Elizabeth Pennisi, *Science*, 2 January 2009, Vol. 323, No. 5910, page 27.

"Declining Coral Calcification on the Great Barrier Reef", by Glenn De'ath et al., *Science*, 2 January 2009, Vol. 323, No. 5910, pages 116–119.

"Ockham's Razor: The Coral Reef Crisis", by Robyn Williams and Charlie Veron, ABC Radio National, 13 December 2009, http://www.abc.net.au/radionational/programs/ockhamsrazor/the-coral-reef-crisis/3094898.

"The 27-Year Decline of Coral Cover on the Great Barrier Reef and its Causes", by Glenn De'ath et al., *PNAS*, 30 October 2012, Vol. 109, No. 44, pages 17995–17999.

"Ocean View: Like a Google Map of the Seabed, a New Survey is Documenting Coral Reefs – Before They're Gone", by Bryan Walsh, *Time*, 14 April 2014, pages 28–33.

"Reef Madness: Dumping Spoil from Port Expansion Projects into the Coral Sea will do Serious Damage to the Great Barrier Reef", by Jon Brodie, *New Scientist*, Vol. 222, No. 2970, 24 May 2014, pages 28–29.

"Ediacaran Metazoan Reefs from the Nama Group, Namibia", by A.M. Penny et al., *Science*, 27 June 2014, Vol. 344, No. 6191, pages 1504–1506.

"Vortical Ciliary Flows Actively Enhance Mass Transport in Reef Corals", by Orr H. Shapiro et al., *PNAS*, 16 September 2014, Vol. 111, No. 37, pages 13391–13396.

"Deep and Complex Ways to Survive Bleaching", by John M. Pandolfi, *Nature*, 5 February 2015, Vol. 518, No. 7537, pages 43–44.

"Predicting Climate-Driven Regime Shifts versus Rebound Potential in Coral Reefs", by Nicholas A.J. Graham et al., *Nature*, 5 February 2015, Vol. 518, No. 7537, pages 94–97.

"Coral Comeback: Reefs Have Secret Weapon against Climate Change", by Michael Slezak, *New Scientist*, 20 June 2015, Vol. 226, No. 3026, pages 36–39.

22 ANTIOXIDANTS & SNAKE OIL

"Antioxidants Prevent Health-Promoting Effects of Physical Exercise in Humans", by Michael Ristow et al., *PNAS*, 26 May 2009, Vol. 106, No. 21, pages 8665–8670.

"Multivitamin/mineral Supplements: Fact Sheet for Health Professionals", National Institutes of Health Office of Dietary Supplements, 7 January 2013, http://ods.od.nih.gov/factsheets/MVMS-HealthProfessional.

"Antioxidants: Not the Superfood Ingredient They Promised to Be", by Dyani Lewis, *ABC Health & Wellbeing*, 1 October 2013, http://www.abc.net.au/health/features/stories/2013/10/01/3859751.htm.

"Healthy Adults Shouldn't Take Vitamin E, Beta Carotene: Expert Panel", by Brenda Goodman, *HealthDay*, 24 February 2014, http://consumer.healthday.com/vitamins-and-nutrition-information-27/beta-carotene-news-57/healthy-adults-should-not-take-vitamin-e-beta-carotene-expert-panel-685178.html.

"The Promise and Perils of Antioxidants for Cancer Patients", by Navdeep S. Chandel and David A. Tuveson, *The New England Journal of Medicine*, 10 July 2014, Vol. 371, No. 2, pages 177–178.

"Fruits and Vegetables Are Trying to Kill You", by Moises Velasquez-Manoff, *Nautilus*, 17 July 2014, http://nautil.us/issue/15/turbulence/fruits-and-vegetables-are-trying-to-kill-you.

"The Antioxidant You're Taking Is Snake Oil", by Tom Philpott, 23 July 2014, *Mother Jones*, http://www.motherjones.com/tom-philpott/2014/07/everything-we-know-about-antioxidants-and-vitamins-wrong.

23 (NOT) NEUTRAL NIGHTLY NEWS

Feedback, *New Scientist*, 6 December 2014, Vol. 224, No. 2998, page 56.

24 HOLE HEARTED

"Mothers' Exercise May Lower Heart Risks In Newborns", by Gretchen Reynolds, *The New York Times* Well blog, 8 April 2015, http://well.blogs.nytimes.com/2015/04/08/exercise-may-lower-heart-risks-in-newborns-study-suggests.

"The Maternal-Age-Associated Risk Of Congenital Heart Disease Is Modifiable", by Claire E. Schulkey et al., *Nature*, 9 April 2015, Vol. 520, No. 7546, pages 230–233.

"Race For Healthy Hearts", by Marc-Phillip Hitz et al., *Nature*, 9 April 2015, Vol. 520, No. 7546, pages 160–161.

25 COWS MAKE FLAVOURED MILK

"Effect of Feed on Flavor in Dairy Foods", by Gerda Urbach, *Journal of Dairy Science*, December 1990, Vol. 73, No. 12, pages 3639–3650.

"Breast Milk Provides Menu of Different Flavours", by Linda Geddes, *New Scientist*, 23 July 2008, Vol. 199, No. 2666, page 14.

"Differential Transfer of Dietary Flavour Compounds into Human Breast Milk", by Helene Hausner et al., *Physiology & Behaviour*, September 2008, Vol. 95, No. 1–2, pages 118–124.

"If You Fed Cows Strawberries, Would It Give Their Milk a Strawberry Flavour?", by Luis Villazon, *BBC Focus Magazine*, March 2015, page 79.

26 COCONUT WATER VS BLOOD

"Clinical Experience in Intravenous Administration of Coconut Water", by B. Eisman et al., *AMA Archives of Surgery*, July 1954, Vol. 69, No. 1, pages 87–93.

"Coco-Nut Water for Intravenous Therapy", by Harry S. Goldsmith, *British Journal of Surgery*, December 1961, Vol. 49, No. 216, pages 29–30.

"The Intravenous Use of Coconut Water", by Darilyn Campbell-Falck et al., *American Journal of Emergency Medicine,* January 2000, Vol. 18, No. 1, pages 108–111.

27 WEIGHT OF A CLOUD

"How Much Does A Cloud Weigh", by Matt Soniak, MentalFloss, http://mentalfloss.com/article/49786/how-much-does-cloud-weigh.

28 VEINS ARE NOT BLUE

"Why Do Veins Appear Blue? A New Look at an Old Question", by Alwin Kienle et al., Applied Optics, 1 March 1996, Vol. 35, No. 7, pages 1151–1160.

29 BACTERIOPHAGE & L'ORÉAL

"Who Discovered Bacteriophage", by Donna H. Duckworth, *Bacteriological Reviews*, December 1976, Vol. 40, No. 4, pages 793–802.

"Set a Bug to Catch a Bug", by John MacGregor, *New Scientist*, 5 April 2003, https://www.newscientist.com/article/mg17823895-100-set-a-bug-to-catch-a-bug.

"Virus Cleans Up Food Poisoning Bug", by James Randerson, *New Scientist*, 23 April, 2003, http://www.newscientist.com/article/dn3652-virus-cleans-up-food-poisoning-bug.

"Bug Killers", by Thomas Häusler, *Nature Medicine*, 1 June 2006, Vol. 12, No. 6, http://www.nature.com/nm/journal/v12/n6/abs/nm0606-600.html.

"The Gene Weavers", by Garry Hamilton, *Nature*, 8 June 2006, Vol. 441, No. 7904, pages 683–685.

"Viruses to Be Sprayed On Cold Meats", *New Scientist*, 27 August 2006, http://www.newscientist.com/article/dn9863-viruses-to-be-sprayed-on-cold-meats.

"Corynebacterium diphtheria: Genome Diversity, Population Structure and Genotyping Perspectives", by Igor Mokrousov, *Infection, Genetics and Evolution*, January 2009, Vol. 9, No. 1, pages 1–15.

"Bacteriophage Therapy: Exploiting Smaller Fleas", by Stan Deresinski, *Clinical Infectious Diseases*, 15 April 2009, Vol. 48, No. 8, pages 1096–1101.

"Viruses Could Kill Superbugs that Antibiotics Can't", by Catherine de Lange, *New Scientist*, 25 April 2009, https://www.newscientist.com/article/mg20227054-500-viruses-could-kill-superbugs-that-antibiotics-cant.

"Fabricating Genetically Engineered High-Power Lithium-Ion Batteries Using Multiple Virus Genes", by Yun Jung Lee et al., *Science*, 22 May 2009, Vol. 324, No. 5930, pages 1051–1055.

"Driller Killers: Turning Bacteria's Weapons on Them", by Roberta Kwok, *New Scientist*, 25 January 2012, https://www.newscientist.com/article/mg21328482-000-driller-killers-turning-bacterias-weapons-on-them.

"Phage Therapy: Concept to Cure", by Eric C. Keen, *Frontiers in Microbiology*, 19 July 2012, http://journal.frontiersin.org/article/10.3389/fmicb.2012.00238/full.

"Rise of the Nano Machines", by Michail C. Roco, *Scientific American*, May 2015, Vol. 308, No. 5, pages 34–35.

"Meet Your New Symbionts: Trillions of Viruses", by Carl Zimmer, *National Geographic* Phenomena blog, 20 May 2013, http://phenomena.nationalgeographic.com/2013/05/20/meet-your-new-symbionts-several-trillion-viruses.

"Bacteriophage Adhering to Mucus Provide a Non–Host-Derived Immunity", by Jeremy J. Barr et al., *PNAS*, 25 June 2013, Vol. 110, No. 26, pages 10771–10776.

"I'm Breeding Biodegradable Batteries from Viruses", *New Scientist*, 26 February 2014, http://www.newscientist.com/article/mg22129580.400-im-breeding-biodegradable-batteries-from-viruses.

"Infections Infected", by Linda Marsa, *Discover*, March 2014, pages 20–22.

"The Age of the Phage", by Shigenobu Matsuzaki et al., *Nature*, 1 May 2014, Vol. 509, No. 7948, page S9.

"Phage Therapy Gets Revitalized", by Sara Reardon, *Nature*, 4 June 2014, Vol. 510, No. 7503, pages 15–16.

"Elizabeth Taylor, My Great-Grandpa, and the Future of Antibiotics", by Michael White, *Pacific Standard*, 27 June 2014, http://www.psmag.com/nature-and-technology/elizabeth-taylor-great-grandpa-future-antibiotics-84626.

"The CD27L and CTP1L Endolysins Targeting *Clostridia* Contain a Built-in Trigger and Release Factor", by Matthew Dunne et al., *PLoS Pathogens*, 24 July 2014, e1004228.

"How to Develop New Antibiotics", by Ezekiel J. Emanuel, *The New York Times*, 24 February 2015. http://www.nytimes.com/2015/02/24/opinion/how-to-develop-new-antibiotics.html.

30 EARWAX & ARMPIT SWEAT

"Anatomy and Orientation of the Human External Ear", by Lynn S. Alvord and Brenda L. Farmer, *Journal of the American Academy of Audiology*, December 1997, Vol. 8, No. 6, pages 383–390.

"Physiology, Pathophysiology, and Anthropology/Epidemiology of Human Ear-canal Secretions", by Ross J. Roeser and Bopanna B. Ballachanda, *Journal of the American Academy of Audiology,* December 1997, Vol. 8, No. 6, pages 391–400.

"The Active Earcanal", by Robert J. Oliveira, *Journal of the American Academy of Audiology*, December 1997, Vol. 8, No. 6, pages 401–410.

"Scientists Find Gene that Controls Type of Earwax in People", by Nicholas Wade, *The New York Times*, 30 January 2006.

"A SNP in the ABCC11 Gene Is the Determinant of Human Earwax Type", by Koh-Ichiro Yoshiura et al., *Nature Genetics*, March 2006, Vol. 38, No. 3, pages 324–330.

"People without Gene for Underarm Odour Still Wear Deodorant", by Tia Ghose and LiveScience, *Scientific American*, 17 January 2013, http://www.scientificamerican.com/article/people-without-underarm-protection.

"The Pharmacogenetics of Body Odor: As Easy as ABCC?", by Sara Brown, *Journal of Investigative Dermatology*, July 2013, Vol. 133, No. 7, pages 1709–1711.

"Dependence of Deodorant Usage on ABCC11 Genotype: Scope for Personalized Genetics in Personal Hygiene", by Santiago Rodriguez et al., *Journal of Investigative Dermatology*, July 2013, Vol. 133, No. 7, pages 1760–1767.

31 MICE ON WHEELS

"Behavior of Captive White-Footed Mice", by J. Lee Kavanau, *Science*, 31 March 1967, Vol. 155, No. 3770, pages 1623–1639.

"Voluntary Wheel Running: A Review and Novel Interpretation", by C.M. Sherwin, *Animal Behaviour*, July 1988, Vol. 56, No. 1, pages 11–27.

"When Wild Animals Encounter an Exercise Wheel", by Jennifer Viegas, *ABC Science*, 21 May 2014, http://www.abc.net.au/science/articles/2014/05/21/4009119.htm.

"Mice, Other Animals Hop onto Exercise Wheel Even On The Wild", by Rhodi Lee, *Tech Times*, 22 May 2015, http://www.techtimes.com/articles/7335/20140522/mice-other-animals-hop-onto-exercise-wheel-even-on-the-wild.htm.

"Wild Mice Run for Fun on Wheels", *Nature*, 29 May 2015, Vol. 509, No. 7502, page 537.

"Wheel Running in the Wild", by Johanna H. Meijer and Yuri Robbers, *Proceedings of the Royal Society B*, 7 July 2014, Vol. 281, No. 1786.

32 TOILET SEAT GAVE SOUND RANGING

"Lawrence Bragg's Role in the Development of Sound-Ranging in World War I", by William Van Kloot, Notes & Records of the Royal Society, 22 September 2005, pages 273-284.

33 ANAESTHETICS DOSE

"A Conserved Behavioural State Barrier Impedes Transitions Between Anesthetic-Induced Unconsciousness and Wakefulness: Evidence for Neural Inertia", by Eliot B. Friedman et al., *PLoS One*, July 2010, Vol. 5, No. 7, e11903.

"What Doctors Don't Understand About Anesthesia", by Stephen Dougherty, *Scientific American*, 28 February 2012, http://www.scientificamerican.com/article/anesthesia-what-doctors-dont-understand.

"The Anesthesia Dilemma", by Dina Fine Maron, 30 June 2015, *Scientific American*, http://www.scientificamerican.com/article/the-anesthesia-dilemma.

34 EXPENSIVE TEACUP

"Collector Gets 422 Million AmEx Points with Cup Purchase", by Frederik Balfour, *Bloomberg Business*, 19 July 2014.

"Five Helpful Suggestions for the Chinese Man Who Suddenly Finds Himself with Millions of AmEx Points to Burn", by Sarah Larimer, *The Washington Post* WorldViews blog, 26 July 2014, http://www.washingtonpost.com/blogs/world-views/wp/2014/07/26/five-helpful-suggestions-for-the-chinese-man-who-suddenly-finds-himself-with-millions-of-amex-points-to-burn.

"At \$38m, the World's Most Expensive Tea Cup", *The Australian Financial Review*, 31 July 2014.

35 FAT: WHERE DOES IT GO?

"Regional Differences in Cellular Mechanisms of Adipose Tissue Gain with Over-feeding", by Yourka D. Tchoukalova et al., *PNAS*, 19 October 2010, Vol. 107, No. 42, pages 18226–18231.

"When Somebody Loses Weight, Where Does the Fat Go?", by Ruben Meerman and Andrew J. Brown, *British Medical Journal*, 16 December 2014, Vol. 349, No. 7988, g7257.

36 EARTH'S SPIN AXIS IS SHIFTING

"The Excitation of the Chandler Wobble", by Richard S. Gross, *Geophysical Research Letters*, 1 August 2000, Vol. 27, No. 15, pages 2329–2332.

"A Second Chance To Save The Climate", by Michael Marshall, *New Scientist*, 22 May 2013, Vol. 218, No. 2918, pages 8–9.

"Rapid Ice Melting Drives Earth's Pole to the East", by J.L. Chen et al., *Geophysical Research Letters*, 16 June 2013, Vol. 40, No. 11, pages 2625–2630.

"Earth's Poles are Shifting Because of Climate Change", by Anil Ananthaswamy, *New Scientist*, 17 December 2013, Vol. 220, No. 2948, page 12.

"Major Antarctic Ice Survey Reveals Dramatic Melting", by Michael Slezak, *New Scientist*, 26 March 2015, http://www.newscientist.com/article/mg22630153.700-major-antarctic-ice-survey-reveals-dramatic-melting.html.

"Volume Loss from Antarctic Ice Shelves Is Accelerating", by Fernando S. Paolo et al., *Science*, 17 April 2015, Vol. 348, No. 6232, pages 327–331.

37 BOUNCING BATTERIES

"In Situ Microtomographic Monitoring of Discharging Processes in Alkaline Cells", by A. Haibel et al., *Journal of the Electrochemical Society*, January 2010, Vol. 157, No. 4, pages A387–A391.

"The Relationship between Coefficient of Restitution and State of Charge of Zinc Alkaline Primary LR6 Batteries", by Shoham Bhadra et al., *Journal of Materials Chemistry A*, 2015, Vol. 3, No. 18, pages 9395–9400, DOI: 10.1039/c5ta01576f.

ACKNOWL-EDGEMENTS

I'M SOMEONE WHO LOVES A GOOD CLICHÉ – THIS SEEMS COUNTERINTUITIVE, GIVEN THAT I'M A WRITER. YOU SEE, WHILE A CLICHÉ IS A PHRASE THAT HAS BECOME OVERUSED – AT SOME STAGE, IT MUST HAVE RUNG TRUE. AND THAT INSIGHT IS WHAT PERSISTS, AND WHAT I'M ATTRACTED TO.

... Which is my excuse for starting these Acknowledgements with a cliché.

So like many others before me, I would like to thank my family – without whom, this book would have been literally (not metaphorically) possible.

Mary knows all she does for me. And my girls, Alice and Lola, are way better than me at punchlines.

In Publishing Land, let me thank Claire Craig (publisher), Jo Butler (agent), Libby Turner, Sarah Fletcher and Emma Rafferty (I need a whole editing team to get the stories ship-shape – another cliché), Alex Christie and Charlotte Ree (publicity), and Jon MacDonald and Roy Chen at Xou Creative.

I also thank Isabelle Benton (Producer, University of Sydney) and Dan Driscoll and Tiger Webb (Producers, Australian Broadcasting Corporation). They all helped craft and shape the stories just after they were born – and also give lovely punchlines.

I love that Books are a manifestation of the Light of Knowledge.

As a Science Journalist, I find myself in the wonderful position of being able to write about fascinating research. Reading the scientific literature gets me a certain level of understanding – but there's all this extra stuff that the scientists know, yet never write down.

I have been lucky enough to get some of this usually unwritten info straight from the horse's mouth, by getting experts in various fields to check my written material. So I thank, in alphabetical story order the following:

Dr. Michael Dobbie, "Anaesthetic Dose";

Dr. Kim Bell-Anderson, "Coconut Oil and Coconut Water" and "Saturated Fats";

Alexander R. Klotz (Physics student), "Fall Through the Earth";

Ruben Meerman (physicist, reading for Ph. D.), "Fat, Where Does it Go?";

Professor Ove Hoegh-Guldeberg, "Great Barrier Reef";

Darryn Schneider, "Naked in Antarctica – the 300 Club";

Professor Geraint Lewis and Professor Zdenka Kunic, "Quantum Life";

Professor Dale Bailey, "Radioactive You";

As always – you all get credit for the good stuff, and I accept the blame for any errors, incorrect slants and misinterpretations.

ALSO BY
DR KARL KRUSZELNICKI

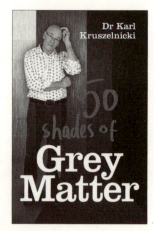
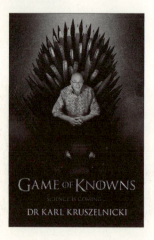
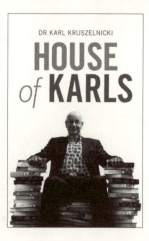